HOSTA

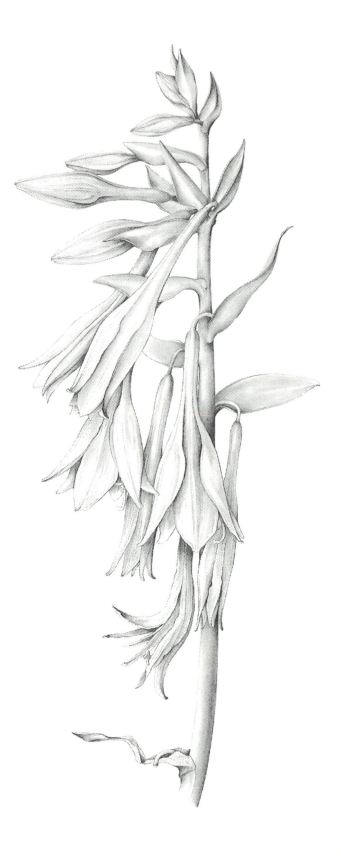

Flower-head

HOSTA
The flowering foliage plant

Diana Grenfell

with drawings by Jenny Brasier

B.T. Batsford Ltd · London

© Diana Grenfell 1990
First published 1990

All rights reserved. No part of this publication
may be reproduced, in any form or by any
means, without permission from the Publisher

ISBN 0 7134 5072 X

Typeset by
Lasertext Ltd.,
Thomas Street,
Stretford,
Manchester M32 0JT

and printed by
Courier International,
Tiptree, Essex

for the Publisher
B.T. Batsford Ltd
4 Fitzhardinge Street
London
W1H 0AH

For Claudia and Colin

CONTENTS

FOREWORD
by C.D. Brickell

The genus *Hosta*, commonly known as the funkia or plantain lily, has risen rapidly from relative obscurity to a zenith of popularity within a few decades, and the current influx of new cultivars, largely from the United States, suggests that within a further decade or two the same problems of distinguishing between closely similar cultivars may beset them as has occurred with daffodils, day-lilies and many other groups of popular garden plants.

With hostas the situation is even more complex due to the uncertain taxonomic position of many of the plants grown as species, some of which are undoubtedly of garden origin. No modern taxonomic revision of the genus as a whole has been published since the Japanese botanist Maekawa produced his monograph in 1940 with a further useful, if controversial, partial study by Fujita in 1976. Regrettably a generally accepted classification of the genus is still awaited.

Happily, Diana Grenfell has now produced for gardeners – and for botanists also – this excellent horticultural account of the genus based not only on her very considerable personal knowledge of hostas but a detailed study of the available literature combined with much research into the taxonomic complexities of this often bewildering group of garden plants. She does not in any way claim that her work is a taxonomic monograph but undoubtedly her extremely well-conceived review of the species and their variants, which forms a major part of the book, will ensure that it will remain as a standard reference point for years to come.

In addition, she has included detailed descriptions of her selection of the cream of the cultivars available, providing a most useful guide to assist the non-specialist who may be confused by the choice of hostas now offered in the nursery trade. The text is complemented by Jenny Brasier's elegant and informative pencil drawings.

We have waited long for a work of high quality and accuracy on the genus *Hosta* to appear. Diana Grenfell in this book has achieved what some believed to be unachievable – an account which most successfully blends a reasoned approach to the taxonomy of the genus combined with authoritative details of *hosta* cultivation and their uses in the garden. She is greatly to be commended on the quality of this book and there is little doubt that *Hosta: The Flowering Foliage Plant* will be found by all gardeners interested in herbaceous plants to be an invaluable aid to their enjoyment.

C.D. Brickell, 1990

ACKNOWLEDGEMENTS

My thanks are due to many people in many parts of the world, for their help with this book.

Firstly I should like to thank Chris Brickell VMH, Director General of the Royal Horticultural Society and President of the British Hosta & Hemerocallis Society, who did more than anyone to create the current interest in hostas. His encouragement and help was freely given and gratefully accepted.

I should also like to thank Jenny Brasier who made the drawings for this book. Her keenly observed illustrations combine accuracy with beauty.

My thanks also to George Schmid for allowing me access to information he has researched over many years, having first learned Japanese in order to glean valuable facts about hostas never before available to western students of the genus; to Dr Warren Pollock, former Chairman of the American Hosta Society's Nomenclature Committee and former editor of the American Hosta Society's bulletin, *The Hosta Journal*, for keeping me in touch with the hosta scene in the United States; to Peter Ruh for much useful information (particularly on the Tardiana Group); to Ronald Smith for the wealth of information about his brother, Eric, and for giving me Eric's own personal hosta slides; to Constance Williams for family history concerning her mother, Mrs Frances Williams; to Alex Summers for sending me hostas long before they became available in Britain; to Jim Archibald for his recollections of The Plantsmen Nursery and of his partnership with Eric Smith, and for his observations on many hostas of the Tardiana Group.

I should also like to thank two immensely helpful librarians: Dr Brent Elliott of the Royal Horticultural Society's Lindley Library and Dr Peter Brunet, Hon. Librarian of the British Hosta & Hemerocallis Society. My thanks also to Piers Trehane for print-outs of the names and authors he has now adopted in his invaluable *Index Hortensis*. Also to Jim Keesing of the Royal Botanic Garden, Kew for information on hostas at Kew.

My thanks also to Julian Grenfell for supplying me with slides of hostas and other photographic materials, and to Dick Kitchingman whose idea it was to found the British Hosta & Hemerocallis Society and who put his notes and card-index systems at my disposal.

Thanks are also due to Doreen Stevens for supplying me with much useful cultural information; to Ann and Roger Bowden, Hon. Registrar and Hon. Secretary respectively of the British Hosta & Hemerocallis Society for their help in the identification of some lesser-known hostas of the Tardiana Group, and in the registration of hostas; to Alan Eason for information on Eric Smith's work at Hadspen House.

I should also like to thank members of the staff of the Royal Horticultural Society's garden at Wisley: Russell Cotes, Andrew Hart, Andrew Halstead, Pippa Greenwood.

My thanks reach to continental Europe – in particular to Ignace van Doorslaer who spent time researching information for me at several university libraries on the three doctor/botanists, Kaempfer, Thunberg and Siebold. He arranged not only my introduction to Dr Karel Hensen who kindly gave me a guided tour of the hostas growing at Wageningen Agricultural University, but also my introduction to Jelena and Robert de Belder at whose estate I was able to see huge, mature clumps of many of Siebold's hosta introductions. Thanks also to Dr Ullrich Fischer for information on German nurserymen.

I should like to thank, too, Jean Sambrook, Secretary of the Hardy Plant Society for giving me much useful advice on the use of hostas for flower arrangement; Anita Hobson, Vice-President of NAFAS (Wessex) and Pam McNicol, President of NAFAS (Wessex), for their kindness in providing background information on NAFAS and the flower-arranging movement; George Smith for valuable information on which hostas to grow for flower arrangement; Sandra Bond, of Goldbrook Plants, for information on how to grow and prepare hostas for exhibition. This has been supplemented by published information by Allen C. Haskell.

I owe particular thanks also to John Bond VMH, The Keeper of the Gardens, Windsor Great Park who, on many occasions, has supplied information on hostas growing in the Savill and Valley Gardens; and to Ken Burras, former Curator of Oxford Botanic Gardens, for background information on Mrs Williams and *Hosta* 'Frances Williams'.

My thanks are also due to the National Hosta Reference Collection holders, The Royal Horticultural Society; Lt Col. and Mrs Jordan; Terry Exley, Horticultural Officer, Leeds Parks Department, and to Michael Heger, formerly of Minnesota Arboretum; and to Liz and Harold Read, Diana and Peter Chappell, David Mason (John Lewis Partnership), Dr Dilys and Dr Tudor Davies, Paul Hobhouse, Ali and Warren Pollock, Christopher Lloyd VMH, and Adrian Bloom VMH, for allowing me to write about their gardens.

I am exceedingly grateful to Hajime Sugita, who kindly translated Kenji Watanabe's book on hostas into English and gave me much valuable information on the growing of hostas in Japan; to Kenzo Watanabe for information on hostas (particularly the *H. montana* species) he collected in the wild with his father.

My sincere thanks to Professor Samuel Jones and Barry Yinger for information on the new hostas from Korea.

And last, but in no way least, my thanks to Roger, who has shared with me the trials and tribulations of writing this book and who has given me so much help, advice and constructive criticism.

Diana Grenfell

LIST OF
ILLUSTRATIONS

Drawings *referred to as 'Fig. oo', etc. in the text*

All drawings by Jenny Brasier, approximately 50% life size and including a petiole cross-section.

INTRODUCTION

It is always tempting to claim that one is writing the definitive book which will appeal to beginner and expert alike. This I have not tried to do: it would be impossible anyway. The novice is now well-catered for in countless hosta articles in the gardening press. Botanists and taxonomists, on the other hand, will continue to re-classify the genus for many years to come in view of the new cytological information now becoming available.

What is needed, I believe, is a book for those whose interest in the genus has already been awakened, who need to know about hostas now and who may not wish to wait a decade or more until the taxonomists have completed their researches.

I have been studying the genus *Hosta* for over 25 years. As a young gardener I was attracted by the sumptuous leaves of some of the larger taxa and their potential as superbly architectural foliage plants. Soon I determined to learn more about them. This proved difficult as there was no comprehensive book on the subject in any language. I began to study the few monographs then available. The major work was written in Japan during World War II, when contact with the West was out of the question and the flow of information was seriously curtailed. The other important monograph was written by a European, who had no first-hand knowledge of the Japanese species and who relied wholly on plants then growing in Europe.

Though my main aim has been to write a book for gardeners and collectors, the genus *Hosta* is so beset with problems of nomenclature and identity that any serious student of the subject must come to terms with these problems if he is to grow the plant he intends.

Since these problems have historical origins I have devoted a whole chapter to the early European collectors and the history of hostas, and another to the botanists, taxonomists and specialist collectors who have contributed to our knowledge of hostas.

In the chapter on species I have attempted to resolve such nomenclatural problems as I can, and to be as accurate and up to date as possible: but in the world of nomenclature and taxonomy nothing is final and no doubt many names will change in the future as a result of new knowledge.

The chapter on cultivars is selective, not exhaustive. I have included only those cultivars which seem to me, from experience and observation, both here and in other countries, to be good garden plants. I see no point in listing every hosta ever registered, since many were not gardenworthy and in any case have already disappeared. The most important thing is to ensure that none of the species ever disappears from cultivation.

Rather than trying to lay down the law as to where hostas should be used in the garden, I have included short accounts of their uses in several gardens to show the diversity of garden habitats to which they lend themselves, but for easy reference I have listed brief tables of the categories into which they seem to fit – again, only including those I believe to be the best of their particular category.

I have looked at the cultivation of hostas and the problems of the pests and diseases which afflict them.

Finally, I have tried to take due cognizance of the enormous geographic area over which hostas are regarded as worthy of garden cultivation and of the varying conditions of climate and light. Plainly and inevitably most of my observations have been made in my own gardens but I have visited a number of gardens in America and Europe to study how hostas are grown in other countries. I have also corresponded lengthily with Japanese academics and hosta growers and discussed hostas with a leading Japanese nurseryman who collects them in the wild. I have been privileged to have been given valued help and information by leading botanists and taxonomists.

It is my hope, therefore, that *Hosta: The Flowering Foliage Plant* will fill what has been a serious gap in the literature and will be enjoyed by hosta growers worldwide.

Diana Grenfell
Apple Court,
Lymington,
Hampshire,
1990

1.
THE HISTORICAL PERSPECTIVE

The first collectors

From the introduction of the very first hosta there has been confusion as to which name denotes which hosta, and this confusion has increased over the years.

The story begins in Japan at the end of the seventeenth century during the period when, under the Tokugawa Shoguns, it was closed to the outside world. The only people allowed in, apart from some Chinese, were medical men employed by the Dutch East India Company. The Dutch were one of the first European nations to establish a trading relationship with the Japanese and the first Dutchmen landed in 1600. They took over a small trading post at Hirado Island formerly run by the Portuguese, but in 1641 this was transferred to Deshima.

Deshima was an artificial island in Nagasaki harbour. It measured 236 paces by 82 paces, just over 32 acres (13 hectares), and was surrounded by a high fence. It was linked to the mainland by a narrow, closely-guarded footbridge. The permitted establishment was a chief factor or director, (*opperhoofd*), a deputy factor, a medical man, a few clerks and some craftsmen, as well as some 200 Japanese servants and interpreters who acted as spies for their Japanese overlords, although many were surprisingly loyal to their European employers even though punishments were barbaric. The tedium of life on Deshima was relieved by the arrival of two Dutch merchantmen each year, and by the visits of prostitutes, the only Japanese women allowed to fraternize with the foreigners.

Engelbert Kaempfer (1651–1716)

The first of three famous medical men on Deshima was Kaempfer. He was a Westphalian German, who later rose to be chief physician to the Dutch fleet. In common with Thunberg and Siebold who followed him, Kaempfer was a shrewd, well-educated man who was deeply interested in plants and took every opportunity to study such Japanese plants as came his way.

The third son of a Lutheran pastor, Engelbert Kaempfer was born in the village of Lemgo in the duchy of Lippe, Westphalia and was 39 by the time he reached Deshima. He had studied medicine, surgery and natural history for three years at Krakow in Poland and continued these studies for four more at Konigsberg in East Prussia. After qualifying in 1680 he made for Uppsala in Sweden, renowned for its science faculty. There he studied for a year under Rudbeck, the first in a long line of eminent Swedish botanists; he then learned that King Charles XI was planning a mission to Persia, and gained an appointment as secretary to the mission under the leadership of a Hollander, Ambassador Ludwig Fabricius. The mission was detained in Russia on the way and, during this time, Kaempfer's passion for collecting and thirst for knowledge manifested themselves, to such an extent that he was thought to have pilfered letters or raided personal files. His ability to learn foreign languages helped him to glean information from many sources. They eventually reached Persia, where he practised medicine and studied Persian culture.

In 1689 he was appointed Chief Surgeon to the Dutch fleet, then in the Persian Gulf, and set off for the Far East. On landing at Batavia he began to study the flora of the island and then, encouraged by his seniors, studied everything he could about Japan. He was urged, when he arrived in Japan, to collect maps, coins, books and papers and send them back to Batavia as his predecessor had done. In May 1690 he sailed for Nagasaki.

Despite the severe restrictions of life on Deshima and his own inability to speak Japanese, Kaempfer's ingenuity and determination soon gained him a foothold in the local community. He exploited two Japanese weaknesses; their inability to hold liquor, and their longing for the forbidden fruit of western knowledge. He taught his Japanese interpreters astronomy and mathematics in exchange for botanical specimens smuggled in from the mainland, even though both he and his interpreters would almost certainly have been put to death had they been caught or betrayed. He started a botanic garden on Deshima, the remains of which are still there.

Kaempfer had the good fortune to join the *Hofreis* or obligatory journey to Edo (now Tokyo), the crowning event of the year for the Dutch, who were required to give reports on the state of all countries with whom they traded, and on what they knew of the activities of the Portuguese. By tradition they were also required to present costly gifts. Accounts of these journeys conjure up a picture of a long and colourful procession of a few Dutchmen, guarded by more than 200 Japanese, crossing the country by land and sea from Nagasaki to Edo. This might have been a wonderful opportunity for foreigners to learn about Japan, but it is most unlikely that Kaempfer would have been able to collect plants on the way. Almost certainly the two hostas known to him were brought to him on Deshima by servants or interpreters.

Kaempfer was the first European knowingly to see a hosta, though we have no actual record as to where or when (or even which). He was also the first to draw one and the first to publish any account of a hosta. His drawings, together with accounts of the hostas, were published in the *Amoenitatum Exoticarum Politico-Physico-Medicarum Fasciculi V*, 1712.

Some time after Kaempfer's death in 1716, Sir Hans Sloane, founder of the Chelsea Physic Garden, bought from Kaempfer's nephew, Dr Johann Herman Kaempfer, Kaempfer's voluminous notes, drawings, and curiosities. It was fashionable at the time to acquire 'cabinets' or 'collections' and because of his interest in natural history and botany, Sir Hans Sloane became a leading collector and charged his 'agents' to obtain Kaempfer's already renowned collection.

Carl Peter Thunberg (1743–1828)

The next medical man of note to serve at Deshima was Swedish, Carl Peter Thunberg who arrived exactly 80 years after Kaempfer had returned to Amsterdam. The fortunes of the Dutch had changed by then and Deshima was no longer a lucrative trading post but a drain on the Dutch economy.

Thunberg was born and brought up at Jonkopping in Smoland, and learned to be resourceful and hardworking at an early age. He enrolled at Uppsala University in 1761 and obtained a doctorate in medicine. He studied under Linnaeus whom he eventually succeeded at Uppsala Botanic Garden. Linnaeus inspired in Thunberg – as he did in many of his pupils – a desire to study the plants of other countries, and in August 1770, Thunberg set out from Uppsala with a travelling scholarship which eventually took him to South Africa, Java and Japan.

His stay in Japan came about as the result of a chance meeting in Amsterdam. There he met Professor Nicholaas Laurens Burmann, a friend of Linnaeus. At that time the wealthy burghers of Amsterdam were spending fortunes introducing exotic fruits and plants for their gardens, and Burmann persuaded them that if Thunberg were to visit Japan on their behalf many good things would come their way. They readily agreed, but thought Thunberg should first have a period of study in the great medical and surgical clinics of Paris. This experience considerably widened his horizons and on his return to Amsterdam he was given a commission as surgeon extraordinary in the Dutch East India Company. Thus was born the first syndicate for the introduction of exotic plants.

Because he needed to know and speak Dutch, Thunberg spent three years at the Cape of Good Hope studying botany, geography, ethnology and linguistics. Thus prepared, he reached Japan in August 1775. He remained until November 1776 – a mere 27 months.

During that time, in spite of the restrictions imposed by the Japanese, he collected specimens of some 900 Japanese plants. His methods were somewhat different from those of Kaempfer. The settlement on Deshima was allowed to keep calves, sheep, oxen, hogs, deer and goats; but, there being no grazing on the island, the animals were all kept in stalls. Fodder gathered from the neighbourhood of Nagasaki by servants was brought in twice each day.

Thunberg sifted through this fodder for plants for his herbarium. Many of our garden plants, possibly even some hostas, were first described by Thunberg from these specimens saved from the fodder. They were usually crushed and bruised, often torn and tattered, but these, pressed, are the lectotypes of many now familiar plants. Before long Thunberg was fulfilling his mission to the burghers of Amsterdam and regularly dispatching botanical specimens to them on East Indiamen calling at Nagasaki.

The only times Thunberg is known to have left Deshima are when he was permitted to botanize in Nagasaki and when it was time for the prescribed *Hofreis*. His account of the *Hofreis* is the most interesting of those written by European doctors and the journey was the highlight of his stay in Japan. Still accompanied by a large retinue of servants, the small band of Europeans travelled in relative comfort in covered vehicles, a great improvement on the horseback journey of Kaempfer's time. Thunberg seized every opportunity to gather botanical specimens and was frequently assisted by the Japanese. He even managed to buy plants in Osaka with the connivance of his guards.

Thunberg could not be persuaded to remain in Japan for a second term as he had massive collections of specimens to classify and he wished to resume his academic career. He returned to Amsterdam in October 1778 and renewed his acquaintance with the men who had sponsored his journey. Some of his introductions from Africa and Japan were already planted and blooming at Haarlem.

Thunberg refused a professorship offered by the Dutch because he was anxious to travel to London to study Kaempfer's material. This had been bought by the British Government in 1753 along with the rest of Sir Hans Sloane's collections and formed part of the nucleus of the British Museum, founded in 1754 (where it can still be consulted). He finally returned to Sweden in March 1779 and settled down to classifying his own material having seen Kaempfer's drawings.

One of the burghers who had sponsored his journey to Japan was Martinus Houttuyn (1720–98). Houttuyn (of *Houttuynia* fame), was a collector and merchant of natural curiosities as well as the author of a vast natural history, the *Natuurlyke Historie*, of which 14 volumes, forming 'Deel II' (1773–85) are devoted to plants. Botanists have tended to ignore Houttuyn's works, so that some of the plant names he published have been overlooked.

Houttuyn had already published descriptions of many of Thunberg's plants by 1779 when Thunberg returned to Sweden to start work on his botanical material. This was quite legitimate. Thunberg and Houttuyn were the best of friends and Thunberg had sent Houttuyn duplicates of many of his dried specimens, all meticulously labelled. In spite of this, somewhere between Thunberg and Houttuyn, confusions did arise, which are examined in the next chapter.

In 1784 Thunberg became professor of medicine and botany at Uppsala; he kept up his links with Japan by writing to his former students, despite the Japanese ban against such correspondence.

Phillip von Siebold (1796–1866)

It was the third of the medico-botanists (an ophthalmic surgeon), Phillip Franz Balthasar von Siebold, who achieved the most; he actually brought hostas back to Europe. Siebold was born in Würzburg, Franconia, Southern Germany. Family influences determined that he should follow a medical career and he qualified as a doctor of medicine, surgery and midwifery in 1820 and started to practise at Heidingsfeld, but soon turned to Holland to widen his horizons.

After a spell as court physician he was appointed the youngest of four surgeon majors for the East Indies Army. In February 1823 he arrived in Batavia. Just then the Dutch authorities were looking for an outstanding young man to further their interests in Japan. Siebold was summoned to the Governor General's estate at Buitenzorg which, as an acclimatization garden, was to play an important part in the bringing of plants from the Far East to Europe. The Governor General, van der Capellen, was impressed by Siebold's knowledge and within six months Siebold was bound for Japan. By the time he reached Nagasaki harbour on 8 August he had taken the opportunity of studying everything he could find on Japan – its lore, language and natural history, and in particular the *Flora Japonica* of Thunberg and the manuscripts of Kaempfer.

Siebold threw himself into his new life. Not only was he an ophthalmic surgeon in a country where eye diseases were endemic but he also seems to have been a good teacher, and eager students came to Deshima from far and near. The restrictions confining him to Deshima were soon eased sufficiently to allow him off the island to see patients and to teach medicine and natural history. He was also allowed to collect botanical specimens in and around the city. He was thus exceptionally privileged – and far less restricted than either Kaempfer or Siebold.

As he was not paid by the Japanese for his services, his patients and students expressed their gratitude by giving him gifts: scrolls, screens, books, pottery, and lacquer-ware. He also instructed his students to write essays for him on subjects about which he needed information – Japanese geography, ethnography, natural history and medicine.

Siebold added a further dimension to his knowledge of Japan by sharing his life on Deshima with a Japanese girl whom he called O-taki-san (1802–65). As regulations did not allow Siebold to marry her, she registered herself as a prostitute in order to be able to live with him. O-taki-san bore Siebold a daughter, O-Ine (1827–1903).

Siebold acquired the services of a Japanese artist, Kawahara Keiga, who made woodblocks for him, not only of plants but also of people. He was also joined in 1825 by an artist from Batavia, Carl Hubert de Villeneuve, and by a pharmacist, Heinrich Burger. Siebold was thus able to amass not only the most amazing collection of herbarium material, but also illustrations, drawings and prints of the Japanese flora. He discovered, which it seems neither Kaempfer or Thunberg had, that the Japanese had a considerable knowledge of their own plants. They had, for example, not only vernacular Japanese names for them but also learned names, written in Chinese characters (the province of the scholar), which were comparable with our own botanical Latin binomials, and they had a rich literature on their herbs and flora. Siebold did not have to make intrepid journeys to distant and dangerous corners of Japan to find new plants: Nagasaki and its environs abounded with material never seen before by Europeans. There is even a local hosta (*H. chibai*, syn. *H. tibai*), the 'Nagasaki' hosta. Exciting though this material must have been to Siebold, it has caused problems for later researchers of the genus since many of the specimens thought originally to have come from wild sources probably came from gardens, including the gardens of temples and monasteries. Among these are some of the first known variegated hostas once assumed to be wild species (*H. undulata, H. crispula, H. decorata*).

In Kaempfer's and Thunberg's time the *Hofreis* had been mandatory every year, but this had been reduced to every four years by Siebold's time. He made the journey in 1826. It was a long trek of some 600 miles (960 km). He took his two artists in his retinue disguised as servants. Such was Siebold's fame in Japan that he was well-received everywhere. In Edo itself he was even shown something that no European had ever seen before: the great secret map of the coast of Japan. This was then in the possession of the Court Astronomer, Takahashi Sakuzaemon, who had a copy made for Siebold.

Siebold returned to Deshima with over 1,000 botanical specimens and he had both Japanese vernacular and Chinese scholarly names for every one of them. One of the students who helped him was Keisuke Ito, who much later became Professor of Botany at the University of Tokyo.

During his stay in Japan Siebold had been granted permission to restore the botanic garden at Deshima started by Kaempfer. This had fallen into decay since Thunberg's time. Siebold used the garden as a repository for plants he intended to send on to Europe via the acclimatization garden at Buitenzorg. On his return from Edo, knowing that he was due to sail for Batavia shortly, Siebold started to prepare his plants for the journey. When the *Cornelius Houtteman* docked he transferred 80 crates to the ship, mostly containing plants. Among other items crated were forbidden documents, including the copy of the great secret map.

The *Cornelius Houtteman* was scheduled to sail in October but on the night of 18 September, south-western Japan was devastated by a typhoon. The typhoon blew over but it had stranded the *Cornelius Houtteman* which had foundered at Inasa near Nagasaki. It was not refloated until the December. This delay gave the Japanese authorities time to inspect it. They found the secret map and Siebold was arrested as a spy.

The *Cornelius Houtteman* sailed without him in January 1829. It reached Holland via Batavia in July 1829. Out of the thousands of plants Siebold had sent off, only 80 had survived. Among these were some of the first living hostas to reach Europe. Three, *Funkia cucullata, F. lanceolata* and *F.*

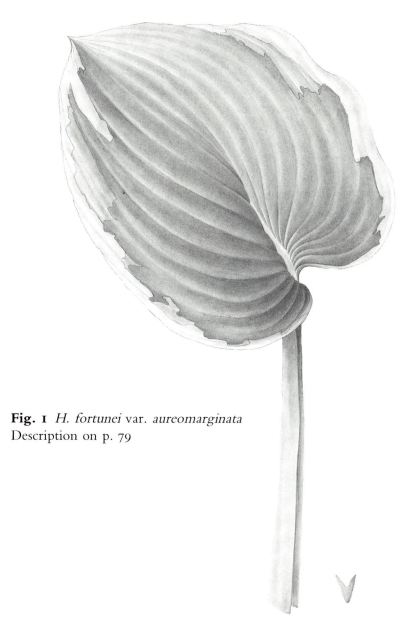

Fig. 1 *H. fortunei* var. *aureomarginata*
Description on p. 79

marginata, found their way to Leyden Botanic Garden probably during 1829.

Meanwhile Siebold had been called before the authorities to answer questions. He steadfastly maintained that he had collected the information and material for scientific purposes. He refused to incriminate his Japanese friends and, after six months, in October 1829, was finally sentenced to be expelled from Japan forthwith and, under pain of death, was barred from ever setting foot on Japanese soil again. He left in disgrace in December 1829. He was 33.

His ship finally arrived at Antwerp on 8 July 1830, and the plants were handed to the curator of the Ghent Botanic Garden on 29 July. Only 260 plants had survived, including some hostas. Records tell us that these were *Funkia maculata, F. spathulata fol. albomarginatis* and, possibly, *F. undulata fol. variegatis.*

Less than a month later, revolution broke out. It was this revolution that led ultimately to the recognition of an independent Belgium. To begin with Siebold had been able to transfer his manuscripts, books and specimens from Antwerp, Brussels and Ghent to Leyden but he could not move his plants, since they were firmly rooted in Belgian soil in Ghent Botanic Garden. According to Siebold, '. . . The best and rarest ones disappeared from the Botanic Garden at Ghent ... several Belgian nurserymen gained honour and profit by growing these plants and the financial result estimated at more than a million francs for the town of Ghent alone'. (Cf. Siebold, 1844; *Krelage*, 1933).

It was to be ten years before Siebold would re-possess any of his own plants from Ghent. A number of Belgian horticulturalists persuaded Ghent Botanic Garden to give him one specimen of each surviving taxa he had introduced in 1830. They supplemented these with plants from their own collections 'as a token of national gratitude for services rendered to Belgian horticulture'. Burger, Siebold's successor at Deshima, also sent bulbs and seeds of Japanese plants among which it is likely that there were hostas.

Meanwhile, Siebold had settled in Leyden. There he set to work writing accounts of his investigations of things Japanese. His collaborator for his botanical works was Professor Joseph Gerhard Zuccharini, a Munich botanist. Their work together culminated in the sumptuously illustrated Siebold & Zuccharini *Flora Japonica*. The first volume, in 20 parts, appeared between 1835 and 1841, and contains 100 colour plates.

Siebold adopted a semi-Japanese life-style in dress and habits which made him rather a pathetic figure. Desperately wanting to see his Japanese plant introductions permanently established in European gardens he purchased a large plot of land at Lagan Rijndik in Leidersdorf. He laid it out in the Japanese manner and called it 'Nippon'. Later, in association with Carl Ludwig Blume, he founded a society for the introduction of Japanese plants and then turned his acclimatization garden into a commercial venture. Many of the catalogues of Siebold & Co. are still preserved at the Dutch Agricultural University at Wageningen. Even the earliest catalogues indicate that hostas featured prominently among the plants offered for sale, although many of the hosta names then used would be unfamiliar to us today. The first catalogue, published in 1848, lists nine species of hosta; in 1856 there were ten species and varieties, and in 1863, 17 species and three varieties. The sudden increase in the number of species offered was due to an unexpected development.

For years Siebold had tried to return to Japan, and in the end the Dutch officially asked the Japanese to lift Siebold's banishment order, so that in 1859, he was allowed to return to Nagasaki to see his daughter. He had last seen her as a child of two; at 31 she was married and the leading midwife in Nagasaki.

The return visit was not a success. He sent back to Siebold and Co. many new introductions, including *Funkia glauca variegata, F. undulata, F. striata, F. angustifolia, F. argento-striato*, and *F. glauca*, which were soon distributed through Europe. The Japanese government invited him to become their adviser on western science and technology, and even on foreign affairs. Siebold also tried to become an unofficial appeaser between Japan and the West. He was the wrong person for such delicate negotiations and succeeded only in antagonizing all sides. He returned to Europe, much embittered, in 1861.

As we shall see, taxonomists and botanists have since tried to classify Siebold's hosta introductions, but with uncertain results. He never published any proper account of his hostas and it would, in many cases, be impossible now to make accurate identifications of Siebold & Co.'s catalogue names, were it not for his herbarium specimens which are lodged at the Botanic Gardens of Leyden and Ghent. These specimens are unfortunately of uneven worth, and many of his unpublished species are missing altogether.

Siebold died in Munich in 1866 and was buried with full military honours. Two of the best known hostas bear his name, *H. sieboldiana* and *H. sieboldii*.

Thomas Hogg

There are four Thomas Hoggs: the father, the son, the nursery and at least one hosta. Oddly there is no published connection between any Thomas Hogg and any hosta.

The name 'Thomas Hogg' is used erroneously in America for *H. decorata*, which was properly described as *H. decorata* by Bailey in 1930, over half a century after its introduction. In Britain the epithet 'Thomas Hogg' is often used for *H. undulata* 'Albo-marginata' and in a looser way for many unidentified white-margined hostas. However, the plant collector, Thomas Hogg jnr, does play a significant part in the history of the introduction of hostas from Japan.

Thomas Hogg snr, started life as a printer but later turned to horticulture. After working for several nurserymen he took charge of the famous gardens of William Kent. He then had premises at Paddington, London, which may have been on the actual site of what is now Paddington Station. He emigrated to the United States in 1820, nine months after his son, Thomas jnr, was born. He arrived in New York intending to go on to Canada but Dr Hosack, founder of the New York City Botanic Garden, persuaded him to stay. He set up as a florist and nurseryman first in Manhattan, New York City, and later took larger premises at Seventy-ninth Street and the East River.

After his death in 1855 the business was run by his two sons, James and Thomas jnr. At that time Thomas was appointed United States Marshall by President Lincoln who sent him to Japan. He arrived in 1862, a few months after Siebold had left for the last time. Robert Fortune may still have been in Japan at that time but there is no suggestion that they met or even knew of each other.

He stayed in Japan for eight years and returned again for a further spell of two years in 1873, this time apparently as an employee of the Japanese Customs Service. He was one of the earliest American travellers and plant collectors in Japan and, as an official representative of the United States, was permitted to go anywhere he liked collecting seeds and plants. He may even have sent back to America hostas identical to those that Siebold had sent to Belgium and Holland. Hogg sent everything he collected back to his brother James at the Thomas Hogg nursery in New York, making the Hogg Collection of Japanese plants the best in the country and probably well-known in Europe as well.

There is no actual record that hostas were sent to the nursery but in all likelihood they were. However, Professor Stephen F. Hamblin, writing on hostas in the 27 June 1936 issue of the *Lexington leaflets* of the Lexington (Mass.) Botanic Garden, says on pp. 211–24, 'there is a plant very similar in bloom to *H. fortunei* (presumably *H. tokudama*) but the leaves are deep green (not at all blue) with the edge decidedly white. This is called *Funkia* 'Thomas Hogg', as it was sent to Dr George Thurber in Passaic, New Jersey, by Thomas Hogg from Japan about 1884. Its interest is in the way the clean white border is put on the leaf margin'. Hamblin's date of 1884 must be wrong, since Hogg left Japan for the second and last time in 1875. The description, however, fits what Bailey published as *H. decorata*, six years after Hamblin's article was published, so this is obviously the source of the original confusion over the names of this hosta. What is not known is why two hostas, namely *H. decorata* and *H. undulata* 'Albo-marginata' (in Britain), both carry the Thomas Hogg epithet. There is no documented evidence that Thomas Hogg, jnr, was responsible for introducing the latter into Britain and Europe, though it is possible. With *H. decorata* the name is easier to understand. This hosta would, quite naturally, have been called 'Thomas Hogg's hosta' and would have been distributed quite widely in America at a time when Asiatic plants were considered a tremendous acquisition. The other problem is that anyone growing a white-edged hosta in England would have had no way of knowing whether one being grown in America was the same, or different, and at that time would probably have assumed the former.

Thomas Hogg jnr is known to have travelled quite widely after 1875, visiting Europe several times, his last visit being in 1880. During any of these journeys he could have introduced both hostas to Europe as well as to Britain, but there is no actual record that this happened. It is widely assumed, but unproven, that Hogg gave *H. decorata* to one of his relatives in the horticultural trade on his visit to Britain. It was certainly growing at Kew in 1930 according to Professor Stearn although there is no record of how and when it was obtained.

American hosta specialists (notably Dr Warren Pollock) researching the origin of the Thomas Hogg name are trying to establish that Thomas Hogg imported both *H. decorata* and *H. undulata* 'Albo-marginata' and think that the plants got muddled somewhere along the way. Professor Stearn also says that he saw *H. undulata* growing next to *H. decorata* at Kew so, again, the muddle could have occurred in Britain. This is of course pure speculation.

Robert Fortune

The specific epithet *fortunei* celebrates Robert Fortune, the Scottish botanist and plant-hunter. His fame rests on the journeys he made as a plant collector. He is best remembered for the plants he brought back from China, many of which were subsequently named for him. His Chinese expedition (1843) was made for the Royal Horticultural Society, by whom he was employed as a plant collector. He subsequently made two further plant-hunting trips, both to the Far East, but as an employee of the East India Company. His final plant-collecting journey was to Japan, and was privately financed. He returned from this journey in 1862 and there is a general supposition that he brought back with him, among other plants, the hosta now called *H. fortunei*, as will be seen in the next chapter.

The vital link is the famous Sunningdale Nurseries of Standish & Noble at Bagshot, Surrey, for it was they who acquired the majority of the plants Fortune brought back from Japan. In 1863 they exhibited at the Royal Horticultural Society's Summer Flower Show a hosta which it is presumed Fortune brought back from Japan, a hosta which has never been seen in a wild habitat. Anonymus, 1863: 'RHS The Third Great Show', *Gard. Chron.*, pp. 626, 628 described the plant they exhibited as a Japanese funkia, with glaucous leaves and French white flowers. This is thought to be the same plant that Regel described as having thickly pruinose bluish leaves and white flowers and was probably also brought from Japan to Europe by Siebold.

The hosta exhibited by Standish & Noble in 1863 and the one described by Regel is still known in Holland and Germany today as *Hosta fortunei*. It should correctly be designated *Hosta tokudama*.

Interestingly, Anonymus also mentioned a variegated form of the same plant, and this he called *Funkia fortunei variegata*. This is also still grown in Holland under that name. It is the hosta which Maekawa later called *Hosta tokudama* 'Aureo-nebulosa'.

The other application of *Hosta fortunei* will be discussed in the next chapter.

2.

A SPATE OF NAMES

The original confusions

The first European to draw, indeed to publish an account of, hostas was Engelbert Kaempfer, in 1712. His illustrations appeared in his *Amoenitatum Exoticarum Politico-Physico-Medicarum Fasciculi V*, the only book (other than his diaries) published by him in his own lifetime. He shows two hostas. The first, no. 52, depicts the plant he identified as 'Joksan, vulgo gibboosi Gladiolus Plantaginis folio' (the common hosta with plantain-like leaves). The other, no. 166, shows the plant he calls 'Gibboosi altera' (the other hosta). All of which is unambiguous.

However, in 1780 Carl Thunberg published in the journal *Nova Acta Soc. Sci.*, Uppsala, a paper entitled 'Kaempfer illustratus', in which he aimed to identify some of the plants mentioned by Kaempfer in *Amoenitatum Exoticarum*, enlarge on their descriptions from his own knowledge of Japanese plants, and give them modern binomials according to the system of Linnaeus. It is here that the problems really begin.

Thunberg gave the binomial *Aletris japonica* to the plant Kaempfer had called 'Joksan, vulgo Gibboosi' (Kaempfer's drawing no. 52). Or did he? A comparison between Thunberg's description and Kaempfer's drawing shows them to be quite different plants. Kaempfer's drawing no. 52 shows a hosta, probably of the Helipteroides section, while what Thunberg describes is what we now call *Hosta lancifolia* (which resembles Kaempfer's drawing no. 166, of which Thunberg makes no mention). It seems that Thunberg based his description not on the Kaempfer drawing, but on herbarium material he had himself brought back from Japan and which he mistakenly thought was the same plant. Thunberg then further confused matters by using the specific epithet *japonica* in his *Flora Japonica*, 1784, for the hosta which then became *Hemerocallis japonica*.

Houttuyn nearly created further confusion by publishing the combination *Aletris japonica* in his *Natuurlyke Historie*, 1773–85. Happily this turns out to be the hosta illustrated in Kaempfer's drawing no. 166 (which is *H. lancifolia* [Fig. 6]), which is of course the same plant as that described by Thunberg as *Aletris japonica*.

Others were confused too. The *Index Kewensis* accepts as valid both *Aletris japonica* (Houttuyn) and *Aletris japonica* (Thunb.), but uses them for different species. Others equate these homonyms with yet other species. Redouté, for example, in 1812, equates Thunberg's *Aletris japonica* with *H. plantaginea* in his revision of the genus, while he says of Kaempfer's drawing no. 166, that it is a good illustration of *Hemerocallis coerulea* (originally spelled *caerulea*) (*H. ventricosa*), which of course, is not what the drawing shows.

Sir Joseph Banks meanwhile had published (in 1791) a copperplate engraving of Kaempfer's drawing no. 52 under the name *Hemerocallis japonica*, this combination having already been made both by Thunberg (in his *Flora Japonica* of 1784) and by Murray (in his *System vegetabilium*).

The way out of this confusion seems to be the course proposed by Bailey in 1930, and endorsed by Hylander in his monograph 'The Genus Hosta in Swedish Gardens', published in 1954 in *Acta Horti Bergiani*, Vol. 16. Bailey's proposal is that the plant for which Thunberg intended the name *Aletris japonica*, and the plant on which the description is in fact based, are two different plants. Bailey suggests that Thunberg's own herbarium material should become the lectotype for *Hemerocallis japonica* (Thunb.) (the *Aletris japonica* of Thunberg in *Nova Acta Soc. Sci.*, Uppsala, 1780). In other words, since Thunberg based his description of *Aletris japonica* on his own herbarium material and since that herbarium material is still preserved in the Botanical Museum at Uppsala, it can now become the type of *Aletris japonica*. Thunberg published that description as for Kaempfer's hosta only because he considered the Kaempfer plant the same as his. But this herbarium material is not Kaempfer's 'Joksan, vulgo Gibboosi' (drawing no. 52). It is Kaempfer's 'Gibboosi altera'. In other words, the *Aletris japonica* of Thunberg *Nova Acta Soc. Sci.*, Uppsala, and the *Hemerocallis japonica* of Thunberg's *Flora Japonica* are one and the same plant – they are the *Hosta lancifolia* of Stearn.

Moreover, it seems certain that the material sent by Thunberg from Japan to Houttuyn – and which Houttuyn published as the type figure of his *Aletris japonica* (Houtt.) in Houttuyn 1781, plate LXXXIV fig. 2 – is also the *Hosta lancifolia* of Stearn, which is both the *Hemerocallis japonica* of Thunberg and the *Aletris japonica* of Thunberg. Thus, in the end it seems that both Thunberg and Houttuyn called the same hosta by the same name. Moreover since the name *Aletris japonica* was communicated to Houttuyn by Thunberg it should be cited as *Aletris japonica* (Houtt. ex Thunb.).

There are, however, other plants, or pieces of plants, remaining only as herbarium fragments, which were brought back to Europe by Thunberg. From his notes and descriptions it is plain that he saw far more plants than he preserved. He mentions details of seed and fruit that are authentic, for example, that *H. lancifolia* does not set fruit, but for which he has no herbarium material. From his herbarium material three species emerge: *Hosta sieboldiana; H. lancifolia;* and a new species, which was later identified as *H. undulata*, a description of which was not validly published until Otto

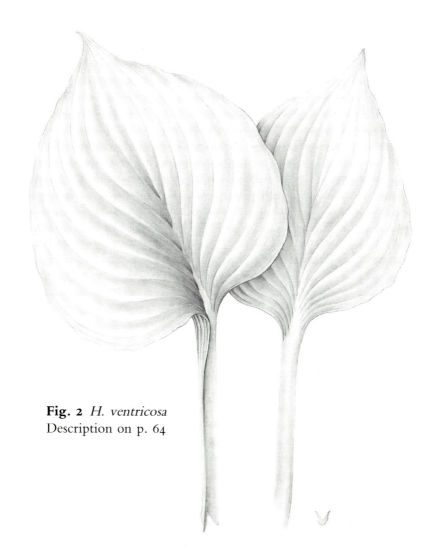

Fig. 2 *H. ventricosa*
Description on p. 64

& Dietrich described the plant from garden material in 1833, as *Funkia undulata* (p. 41), although Thunberg had referred to a *Hemerocallis undulata* in 1797 – without, however, giving any clue as to its identity. Incredible though it may seem, all these confusions had already arisen before a single live hosta even reached Europe.

The first hostas in Europe

The first living hostas to be seen in Europe were raised from seed sent by Charles de Guignes, the French consul at Macao, to the Jardin des Plantes in Paris some time between 1784 and 1789 – 50 years after the publication of Kaempfer's two *giboosi* drawings, nos. 166 and 52. This plant was first described as *Hemerocallis plantaginea* by the French naturalist Lamarck (1789), working from material growing in gardens in Paris. It has since become the typical Parisian hosta. Indeed in *Gard. Chron.*, October 1868, it was said to be cultivated by the thousand in public gardens in Paris, and to be common elsewhere in France. There were said to be long, dense edgings of it near Versailles. Today it is more often seen in France growing in pots and containers in front gardens and doorways as far south as the Midi.

Both *H. plantaginea*, and the other Chinese hosta, *H. ventricosa*, were introduced to Britain in 1790 by the wealthy, self-styled 'promoter of botanical science', George Hibbert (1757–1837). Hibbert was an influential man in the world of business, the arts and horticulture; he was a London alderman, Member of Parliament for Seaford in Sussex and, more interestingly, a West India merchant. He sponsored botanical expeditions and was a member of the 'Original Society', that is, one of the 28 founder members of the Royal Horticultural Society and a Fellow of the Linnaean Society of London as early as 1793. Until 1829 he lived in Clapham, south London where he had his own botanic garden. His gardener was Joseph Knight who later ran the Chelsea Nursery.

Hibbert was also a self-publicist, and it was under his wing that Henry Andrews started in 1797 a rival publication to *Curtis' Botanical Magazine* which, until then, had been alone in its field in publishing superbly illustrated scientific writings. Andrews' *Botanist's Repository* (with plants supplied by Hibbert) limited itself to newly introduced or rare plants and only ten volumes were issued. In its first volume Andrews describes *Hosta ventricosa* (as *Hemerocallis caerulea*), the blue day-lily, from cultivated specimens taken from Hibbert's collection. Andrew states: 'This species, as well as a white variety, a native of China, had been introduced by Hibbert from "thence" [referring to China]' so the presumption is that *H. plantaginea* was brought back for him on one of his merchantmen. Sir Edward Salisbury confirms this when describing *H. plantaginea* as *Niobe cordifolia* (nomen illeg.): '. . . it was introduced in 1790 from China, by George Hibbert Esq., and though quite hardy, the flowers appear so late, as to be often cut off by our autumnal frosts, if it be not planted in a very warm situation. It probably grows wild in moist places, close to the edge of a pond.' No other hosta fits this description.

There is a problem with these seed-raised species in that few people nowadays would consider any hosta raised from seed a true species because hostas are notorious for the freedom with which they hybridize. However, when *H. plantaginea* and *H. ventricosa* were introduced no such objections were raised. To begin with they are the only two hostas native to China: the latter is well-known as the only hosta that comes true from seed – which it produces apomictically (p. 65).

Furthermore, there is no suggestion anywhere that the seed of *H. plantaginea* produced widely differing seedlings – a sure sign of hybridity. Finally, there is no suggestion that Hibbert's introduction, presumably also as seed, differed from that raised in Paris. Indeed it would not have been long before the two introductions would have been growing side by side somewhere, as it is human nature to make comparisons in such circumstances. All the evidence therefore seems to point to the true species having been introduced not once but twice.

This was the age when glass and stove houses were *de rigueur* for all serious plant collectors and wealthy garden owners, and these two hostas were grown extensively under glass in the erroneous belief that they were tender. *H. plantaginea* was often referred to as *Funkia albo*, presumably on account of its white, scented flowers, said to equal those of stephanotis in fragrance. Correspondents to gardening journals of the period on the whole considered that north of London *H. plantaginea* would not flower satisfactorily out-of-doors and that even further north, the leaves were too delicate to bear exposure to wind, even if grown in the shelter of a south wall.

H. plantaginea (the original plantain lily) in its form var. *grandiflora* was later used extensively by Gertrude Jekyll as a pot and tub plant (p. 139), and it is often featured in her books on garden design. She referred to it as *Funkia grandiflora*; she also called other hostas by names that we would not now consider correct. This is not surprising, since *H. plantaginea* has variously been known as *Saussurea subcordata* and *Niobe cordifolia*. *H. plantaginea*, with the exception of the locally grown var. *japonica* (thought to be synonymous with *H. plantaginea* var. *grandiflora*) (p. 64), was only introduced to Japan in the 1930s, although *H.p.* var. *grandiflora* had been widely distributed from Siebold's nursery from the 1860s.

In 1841 *F. grandiflora* Sieb. arrived in Europe from Indonesia. J. Peirot, a member of the staff of the Rijks-herbarium at Leiden, broke his journey to Japan at Indonesia. From there he sent back a selection of plants including some of Japanese origin, presumably planted by Siebold in the botanic garden at Buitenzorg. It is not known if the epithet referred to the type plant or was the form of *H. plantaginea* known as var. *grandiflora*.

The double-flowered form of *H. plantaginea*, known as 'Aphrodite', is of much later introduction and was first mentioned by Dr Maekawa (1940). It was not known outside China until very recently. Some plants of it were sent as a gift from Beijing (Peking) Botanic Garden to the Arnold Arboretum in America.

H. ventricosa was the second hosta to reach Britain and it was imported by Hibbert in 1790, direct from China. What remains a mystery is how he knew that such a plant existed and was worth importing. The only details we have are those given by Andrews in the *Botanical Repository*.

The naming of the genus

When the first hostas reached Europe they were placed in the genus *Hemerocallis*, though before that Thunberg had already named his original plant *Aletris*. Various other names were proposed before the name *Hosta* was finally accepted – if, indeed, the matter is settled even now.

The Chinese and Japanese have their own systems of naming plants, both vulgar and scholarly, so that when in 1712 Kaempfer published his drawings of hostas the most accurate names known were the Japanese vernacular names, which he Latinized as *Joksan vulgo giboosi* and *Altera giboosi*. Such terminology was accepted in the age before Linnaeus had published the binomial System. It was in order to fit Kaempfer's plants into this system that Thunberg gave them the name *Aletris* (*Aletris japonica*, Thunberg 1780). Later (1784), for reasons that are not clear, he transferred them to the genus *Hemerocallis*.

The next revision of the generic name was no doubt prompted by the greater number of species known to Europeans by the beginning of the nineteenth century and by the disparities between them. In 1812 Sir Edward Salisbury proposed not one but two new generic names for the plants we now call hostas, the genus *Niobe* for the plant we know as *Hosta plantaginea* and the genus *Bryocles* for all other hostas. He separated the two genera by a very minor point concerning the bracts on the flowering stem in *Niobe*. Had Salisbury validated these names he would have added a whole further chapter to this comedy of errors, but since he published no description with them these names are without validity.

Hosta Tratt.

By coincidence, 1812 was also the year in which the Austrian botanist, Leopold Trattinick (1764–1848), proposed the generic name *Hosta* to embrace the three species known to him – *Hemerocallis japonica*, H.

Fig. 3 *H. undulata* var. *univittata*
Description on p. 101

lancifolia and *H. caerulea*. Trattinick published his proposed name in Vienna, in *Observ.*, l.c. The name celebrates the Austrian botanist Nicholas Thomas Host (1761–1834), author of a *Flora Austriaca* and a work on grasses; he was also physician to the Emperor Francis II. But Trattinick's name *Hosta* is illegitimate by today's rules because he never formally transferred these three species to his new genus. Further, his name *Hosta japonica* is illegitimate because it appeared without a synonym, though accompanying an engraving. This engraving does not show *Hosta (i.e. Aletris) japonica*: it shows *H. plantaginea*. Trattinick thus inadvertently perpetuates a former confusion. In spite of this the name *Hemerocallis japonica* has been commonly used since it appeared in P.J. Redouté's famous drawing published in 1802. In accordance with Article 75 of the ICBN (International Code for Botanical Nomenclature (1972)) rules it should be completely rejected as a nomen ambiguum.

Trattinick's name *Hosta* is invalid on other grounds. It had already previously been used by Baron N.J. von Jacquin, director of the Botanic Garden in Vienna, for a genus in Verbenaceae now known as *Cornutia*. However, even that use of the name *Hosta* was invalid, since Linnaeus had previously published a valid name for that particular taxa in Verbenaceae. The rule is that once a name has been published for a taxon, even if that publication was invalid, it may not be used again for another taxon. Ultimately, though, the name *Hosta* has been conserved in accordance with Article 14 of the ICBN.

The generic name *Funkia* is more recent than the name Hosta, having been validly published in 1817 in 'Anleitung zur Kenntniss der Gewachse, 2', 1.246 by Kurt Polykarp Joachim Sprengel (1766–1833), Professor of Botany at Halle. In 1825 Sprengel also published the name *Funkia ovata* (*ventricosa*) in *Caroli Linnaei ... Systema Vegetabilium* (Ed. 16a, II, p. 40). *Funckia* (with a 'c') has also been published as a generic name which, again, caused confusion, firstly by Willdenow in 1802, describing the genus now known as *Astelia*, and later in 1818, by Dennstaedt, in association with Schluss, for the genus now known as *Lumnitzera*.

The name *Funkia* has passed virtually unchanged, into most European languages as a vernacular name for plants of the genus *Hosta*. In German it has become 'Funkie'. In choosing the name *Funkia*, Sprengel is thought to have been honouring Heinrich Christian Funk, about whom relatively little is known since he was a practical collector-botanist, rather than a theoretician, and published nothing. He is thought to have been born in 1771, and to have died in 1839.

The disperal of hostas round Europe

In the early 1830s many of the hostas sent to and imported into Belgium and Holland by Siebold were later sent on to England and Germany. One of the first of these, and still one of the best-known, was described and validly published as *Funkia undulata* by Otto & Dietrich in *Allg. gartenztg.* in 1833. This was the first of the variegated species or rather a species of convenience, since that hosta has never been seen in the wild as far as we know.

Loddiges and *H. Sieboldiana*

The hosta we now know as *H. sieboldiana* was sent to England in 1830 from the botanic garden at Leiden; the recipient was the nursery of Loddiges. A Dutchman, Conrad Loddiges (c. 1739–1826), took over the Hackney nursery founded by the German, John Busch, in the early 1770s. This well-known nursery specialized in stove and greenhouse exotics so it seems almost certain that Loddiges supposed this hosta to be a tender subject, as *H. plantaginea* was deemed to be at that period. Loddiges published an account of their hosta without a description in the *Botanical Cabinet* 19, no. 1869, 1832, calling it *Hemerocallis sieboldtiana*.

Six years later (1838) a similar plant was validly described in *Curtis' Botanical Magazine* by William (later Sir William) Hooker. The plant had been sent to him, he said, 'by a Mr. M'Coy', presumably the L.J. Makoy, a Liège nurseryman (Hensen 1963). Hooker states that this plant, *Funckia* (sic) *sieboldiana*, had been introduced to Belgium by Siebold, so it can safely be presumed that it had been imported directly from Japan in the batch of plants which had arrived at Ghent in 1830. The *Hemerocallis sieboldtiana* sent to Loddiges had come from Leiden and was most probably the glaucous form, *H.s. cucullata*. This is borne out by the supposition that Hooker's plant must have had green leaves, as he would undoubtedly have described the leaves as glaucous, had they been so. Loddiges, on the other hand, quite definitely refers to the bluish-green of their plant.

Dr John Lindley also published a description of this plant in 1838 in *Edwards Botanical Register* 25, pl. 50, calling it *Funkia sieboldii*. Moreover, Joseph (later Sir Joseph) Paxton, the Duke of Devonshire's gardener, had also published the combination *Hemerocallis sieboldii* in the same year, but for a quite different species. This duplication of a name obviously causes confusion, yet it stems from the intention of botanists eager to honour Siebold's name. It is understandable that Lindley wished to publish a description – and his is considered the better illustration – since his plant was grown out-of-doors while Hooker's was grown under glass.

The combination *Hosta sieboldiana* was first published by Engler, in Engler and Prantl., *Die naturlichen Pflanzenfamilien*, II Teil, 5, p. 40, in

1888, and its first synonym should therefore be *Hemerocallis sieboldtiana*. The earliest reference is Iinuma in *Somoku Dsusetzu* (Japan) in 1874.

Joseph Paxton

As we saw earlier, Siebold introduced his *Funkia spathulata* fol. *albomarginatis* to Holland in one of his first shipments from Japan. It was one of the common species in the wild in Japan where it is generally a green-leafed plant, though the type plant is variegated. It was subsequently brought to England sometime during the 1830s, probably by Joseph Paxton.

Thus in 1838, just a few weeks apart, both Joseph Paxton and William Hooker published descriptions of this hosta, but by mischance each of them gave it a different name. Paxton called his plant *Hemerocallis sieboldii* and his plate of it appears in the fifth volume of his *Magazine of Botany* No. 1, 25, March 1838. It therefore takes priority by only a matter of weeks over Hooker's *Funckia albomarginata* published on 1 May 1838, in *Curtis' Botanical Magazine*. There has been confusion ever since, because many gardeners use the *sieboldii* epithet in error for the larger-leafed *H. sieboldiana* which, furthermore, was often incorrectly known as *Funckia sieboldii*. This confusion was perpetuated in many gardening publications and nursery catalogues, and even such noted horticulturalists as Gertrude Jekyll and William Robinson used the name *sieboldii* for the wrong plant. It was not until 1942 that the Japanese botanist Jisaburo Ohwi (p. 58) published the specific name *albomarginata* in combination, for the first time, with the generic name *Hosta*. The name *Hosta albomarginata* was considered correct until 1967, when John Ingram of the L.H. Bailey Hortorium, Cornell University, pointed out that under the rules of priority of publication, Hooker's specific epithet *albomarginata* should be replaced with Paxton's *sieboldii*.

The *fortunei* confusion

The *fortunei* epithet, after its first mention in connection with hostas (by Anonymus in 1863, R.H.S. and *Gard. Chron.*), next crops up in the 1870/71 catalogue of Siebold et Cie, where *Funkia fortunei* is offered for sale, as well as *F. fortunei variegata*. The name was not, however, validly published, with a description, until 1876, and then to confuse matters further, it was published twice in the same year by different people, each applying it to a different plant.

In *Gartenflora*, 1876, Regel published a description of a plant he called *Funkia sieboldiana* var. *fortunei*. He based his description on material from plants growing in German gardens. This hosta had probably been distributed by Siebold's nursery who, in their turn, were likely to have received it from

Fig. 4 *H. fortunei* var. *hyacinthina*
Description on p. 78

Standish & Noble of Sunningdale, Berkshire, England. The plant described by Regel appears to be the same as that described by Anonymus (1863 R.H.S. and *Gard. Chron.*) referred to in connection with Robert Fortune, and is, unquestionably, the *H. tokudama* of Maekawa (p. 77).

Unbeknown to Regel, J.G. Baker, then Keeper of the Herbarium at the Royal Botanic Garden, Kew, had earlier in the same year published a description of a hosta which he had also distinguished by the honorific epithet *fortunei*. Quite where the Baker plant came from is not recorded, but it is probable that it may have also been brought back from Japan by Fortune – who, after all, may have collected more than one hosta; this point of view is also propounded by Dr Hylander, the Swedish taxonomist, who made a particular study of the hostas of the *fortunei* complex.

Baker states in *Gard. Chron.*, vol. VI, p. 36, 'I at first hesitated to consider it as more than a variety of *H. sieboldiana* but, after watching it for the last three summers in Mr Barr's garden and having also received tiny specimens from Mr Ware, I am now inclined to look upon it as a species. It is widely spread in gardens under the name I have adopted and it is perfectly hardy'.

There is an authentic specimen of Baker's *Funkia fortunei* in the Kew Herbarium, but it consists of only a small leaf and an inflorescence in a very bad state of preservation. While it is indeed an *H. fortunei* as we now understand it, it is such a poor specimen that it cannot be assigned to any particular clone.

Thus *Funkia sieboldiana* var. *fortunei* Reg. is *F. fortunei* hort. Sieb., and both are *H. tokudama* Maekawa. The equivalences of Baker's *fortunei* are less straightforward. As stated, Baker's *F. fortunei* is represented by an authentic specimen. Bailey examined and indeed photographed this sheet when he was working on the genus, and reproduced part of it as 'part of type sheet' in his *Study*, 'Hosta: the Plantain Lilies' in 1930. However, when Baker was making his description of *F. fortunei*, he attached to this herbarium sheet for purposes of comparison a specimen of *F. sieboldiana* and, by a most extraordinary mistake, the plant that Bailey reproduced was in fact Baker's *F. sieboldiana*, not his *F. fortunei*. However, this muddle over the photograph did not influence Bailey's interpretation of Baker's original *F. fortunei*. The plant Bailey describes is not, after all, Baker's *F. fortunei*, but the plant at that time grown in American gardens under the name *H. fortunei*. Whether or not there is a coincidence of identity is not clear from Bailey's description, which is insufficient for this identification to be made. It is unfortunate that on the one hand there are now no longer any specimens of Baker's *F. fortunei* growing at Kew, and on the other that Bailey left no authentic material for his *H. fortunei* in his herbarium. However, Hylander states that he has seen photographs of plants growing in American gardens, as well as Baker's herbarium material and, from this evidence, he supports the view that Baker's and Bailey's plants are, luckily, the same.

An interesting footnote on the fate of the *fortunei* epithet as Hylander points out, is that were Baker's name to be rejected, the only other validly published binomial would be *H. bella* Wehrh. Hylander rejects Maekawa's idea of ascribing Bailey's plant to *H. montana*, saying merely that Maekawa was mistaken in associating *H. fortunei* with *H. elata*. The view held nowadays by many botanists is that *H. elata* is, in fact a hybrid, with *H. fortunei* as one of its parents (p. 71).

Hosta tardiflora

A long-standing confusion of identity exists between *H. lancifolia* and *H. tardiflora*, which is sometimes listed as a variety of *H. lancifolia*. The first plants of *H. tardiflora* were sent to Kew Gardens by Max Leichtlin (1831–1910) of Baden-Baden in 1885.

H. tardiflora obviously came from Japan although we do not know the route by which it reached Leichtlin and he never published a proper description. He regarded it as a species and named it *Funkia tardiflora*; but in 1902 it appeared in the *Kew Handlist of Herbaceous Plants* as *Funkia lancifolia* var. *tardiflora* and it was also later classified by botanists as a variety of *H. lancifolia*. In 1903 Walter Irving described and illustrated this plant, calling it *Funkia tardiflora*; but there is now some doubt as to whether or not it is, after all, a species. Current Japanese thinking is that it is not.

In 1916 C.H. Wright of the Kew Herbarium wrote a text to accompany the illustration t.8645 in the *Botanical Magazine*, and he preferred to call the plant *Funkia lancifolia* var. *tardiflora*, stating that 'it cannot be distinguished by any good morphological character from *F. lancifolia*.' The plant illustrated had a fasciated scape, giving a misleading impression.

In 1930 L.H. Bailey, when he came to revise the genus, had to accept Wright's view, albeit reluctantly, since he (Bailey) lacked living material. Professor W.T. Stearn in *Curtis' Botanical Magazine*, 1953, set out the differences between *H. lancifolia* and *H. tardiflora*, and published the illustration t.8645 as Irving's *tardiflora*, although in *Gard. Chron.*, 1931, he had shared Bailey's view. However, Maekawa contends that this illustration in fact shows *H. sparsa*, a species described by Professor Nakai from material cultivated near Nagoya province, Owari, Honshu. This material was sent to Kew by Prof. Nakai and is at Kew considered conspecific with *H. tardiflora*, but a different clone. Perhaps the explanation is that the fasciated-scaped specimen noted above is not in fact fasciated *tardiflora* but typical *sparsa*.

H. tardiflora was considered a rarity for the half century following its introduction and very few plants were distributed, although Reginald Farrer, writing in *The English Rock Garden*, 1919, mentions it as 'the only species we have at present fitted in stature for our realm [i.e. the rock garden] where its delicate foot-high spires of lavender pale lilylings in autumn have a fine effect'. Although we now know that there are many small hostas, *H. tardiflora* was, for a long time, one of the few dwarfs and was to have a very significant effect on the development of the genus on account of its late-flowering characteristics (pp. 51 & 123).

Max Leichtlin obviously distributed material of *H. tardiflora* in Europe as well as Britain, since the German nurseryman Georg Arends (1863–1951) of Wuppertal-Ronsdorf in Westphalia states in his memoirs *Mein Leben als gärtner und Züchter* ('My Life as Nurseryman and Breeder') published by Eugen Ulmer, Stuttgart, 1951: 'The species which I had obtained under the name of *Hosta tardiflora* generally blooms so late here [at Ronsdorf] that the flowers [usually] suffered from the first autumn frosts. By crossing forced plants [of *Hosta tardiflora*] with the earlier flowering *Hosta coerulea/ovata* (*H. ventricosa*) I obtained in 1911 *Hosta tardiflora* 'Hybrida' which also starts flowering late, but doesn't suffer from the early autumn frosts'. 'Hybrida' never seems to have been distributed widely but it is now growing in the Trial Grounds at Friesing-Weihenstephan in Bavaria (Faculty of Agriculture, Horticulture and Brewing of the University of Munich). This cross, therefore, pre-dates by approximately half a century Eric Smith's similar but more famous *H.* x *tardiana* hybrids, now correctly known as the Tardiana group.

Georg Arends and *H. sieboldiana* var. *elegans*

Georg Arends is best known for his *Hosta sieboldiana* 'Elegans', still one of the most popular hostas grown today (p. 76). He says: 'From the steel-blue, but a little weak and low-growing *Hosta fortunei* (*H. tokudama* (Maekawa)), I obtained by pollination with the strong-growing blue-green *Hosta sieboldiana* the vigorous *Hosta fortunei* 'Robusta' (later named *Hosta sieboldiana* var. *elegans*) which, with its bluish leaf came close to *Hosta fortunei*. [Close in colour, presumably]. . . . This, I offered for the first time in the year 1905 after selection from many seedlings.'

Arends' *H. fortunei* must be what we now call *H. tokudama* (Maekawa) and not *H. fortunei* (Baker), since he describes the plant as very slow-growing and very blue-leafed. Hylander (1954) states that it has been impossible to settle whether or not the 'Elegans' of Swedish gardens is the same plant as *H. fortunei* 'Robusta'. He thinks that various clones of 'Elegans' exist. This is borne out by photographs and herbarium specimens of American plants which appear indistinguishable from Hylander's var. *elegans*.

Georg Arends was well-known for his work with astilbes, phloxes, azaleas and indoor primulas, but he devoted two whole years to working extensively with hostas. One of his aims was to improve the plant sometimes erroneously called *Hosta minor* (*Hosta sieboldii*) 'Alba', that is, to breed and select a cultivar with larger white flowers. He attempted a cross between *H. sieboldii* 'Alba' and *H. plantaginea* var. *grandiflora* but was unsuccessful, so he then concentrated on producing deeper purple flowers, using *H. ventricosa*.

Improved forms of *H. sieboldii* 'Alba' were in fact raised at Weihenstephan in the 1970s (p. 126). The *H. sieboldii* 'Alba' × *H. plantaginea* var. *grandiflora* cross has since been made successfully and has produced the popular 'Honey-bells' and 'Royal Standard' hybrids.

Hosta rectifolia

The last of the species to be grown in Europe until about the 1950s was *H. rectifolia* and, here again, a major case of misidentification occurred. In the second *Kew Handlist of Herbaceous Plants* (1902), '*Funkia longipes*, Franch. et Sav.' was added to the list of plantain lilies or hostas grown at Kew though, according to Professor Stearn, writing in *Gard. Chron.*, 1932, the true plant of Franchet et Savatier has probably never been introduced from Japan. Stearn goes on to say that whilst at Kew he saw the *Funkia longipes* of the *Kew Handlist of Herbaceous Plants* – and he thought it quite distinct from the genuine *H. longipes* (one of the commonest species in the wild, even accounting for the fact that *H. longipes* has a very wide distribution and range of habitats and is, therefore, quite variable). He sent material to Professor Nakai of Tokyo, who had recently described several new species and took a particular interest in the genus. Nakai at once identified this material as *Hosta rectifolia*, a near relative of *H. decorata*. By an odd chance, the seed from which this plant had been raised had been sent to Kew by Tokyo Botanic Garden in 1879, labelled *H. longipes*. The plants raised from this seed first flowered at Kew in 1901 and were then distributed but were not recognized as a new species until 1930.

This instance is one among many past and present cases of hostas being misidentified because seed has been received wrongly labelled.

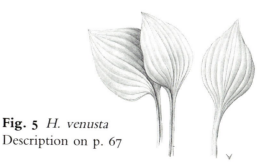

Fig. 5 *H. venusta*
Description on p. 67

Botanists, taxonomists and hosta 'specialists'

One of the earliest attempts to resolve the nomenclatural and taxonomic problems of *Hosta* was that of C.S. Kunth in 1843 in the fourth volume of his *Enumeratio Plantarum* in the section Asphodeleae (Funkia) where he divided the genus into five species, namely (1) *F. subcordata*, (2) *F. ovata*, (3) *F. undulata*, (4) *F. sieboldiana* and (5) *F. albo-marginata*. Carl Sigismund Kunth (1788–1850) was a protégé of von Humboldt, director of the botanic garden at Berlin.

F.A.W. Miquel, professor of botany at Utrecht and director of the Leiden Herbarium, published two revisions of the genus *Funkia*, a preliminary one in 1867 following a study of Siebold's Japanese herbarium material, and another in 1869 after studying living material growing in Siebold's nursery and in the Utrecht botanic garden. Miquel again divided the genus into five species, four of which he put in the subgenus *Giboshi*. He then subdivided the species into varieties and sports, bringing the number of taxa in the genus up to 20. He is best known for his description of *Funkia glauca*, introduced by Siebold in 1862.

Heinrich Witte, curator from 1855 to 1898 of the botanic garden at Leiden, in 1868 named this hosta *F. sieboldiana* var. *glauca*, and Hensen (1963) suggests it may have looked like an intermediate between *H. sieboldiana* var. *sieboldiana* and *H. sieboldiana* var. *elegans*, though of course *H.s.* 'Elegans' was not raised until 1905. Maekawa (1940) claims that it is the same as his *H.s. hypophylla*. The name *glauca* is still sometimes used in a looser sense for almost any glaucous-leafed large hosta.

Miquel's *H. latifolia* puzzled many later interpreters of the genus until it was correctly identified as *H. ventricosa* by Gray on the basis of information from Stearn in 1938. Miquel's most important contribution to the understanding of the genus was the bequest of his herbarium material to the University of Utrecht. This material has been of inestimable value to later botanists and taxonomists.

Many of the garden forms mentioned in the works of Silva Tarouca and Wehrhahn, as well as those of Mottett and de Noter, cannot now be identified so that their studies are not particularly useful to us. George V. Nash attempted to change the name of the genus once again in 1911, only a few years after the name *Hosta* had finally been officially accepted (1905). Nash was at one time head gardener at New York Botanic Garden and his article 'Funkias or Day-lilies' appeared in, *Torreya*, the journal of the Torrey Botanical Club. He proposed the name *Niobe* (Salisbury 1812) on the grounds that *Hosta* was homonymous in spite of the ruling by the International Botanical Congress in Vienna. He accepted Salisbury's division of the genus into the subgroups *Niobe* and *Bryocles*, but was of the opinion that the small bracteole at the base of the pedicel on *H. plantaginea*, the basis for the division, is often not apparent, and so it is better to consider the two groups as one genus. He then divided the genus into six species. This revision is considered by all later students of the genus to be quite incorrect.

Two works of enduring importance from the 1930s are firstly Liberty Hyde Bailey's article in *Gentes Herbarum*, 1930, where he divides the genus into three subgenera, *Niobe* – containing *H. plantaginea*; *Bryocles* – containing *H. ventricosa*; and *Alcyone* – containing all the remaining species; and secondly W.T. Stearn's revision based largely on the work of L.H. Bailey, in *Gard. Chron.*, 1931. Stearn used Bailey as his starting point. To begin with he examined only a few species, but later revised the whole genus as it was known then. It was he who resolved the long-standing problem over the identity of Franchet and Savatier's *H. longipes* (p. 47) revealing it to be *H. rectifolia*.

Just a few years after the Bailey/Stearn contributions, a young Japanese botanist began to study hostas. Fumio Maekawa studied under his father-in-law, Professor Takenoshin Nakai (1882–1952) at the Faculty of Science, Tokyo Imperial University and wrote a monograph on hostas to gain his doctorate. Maekawa little realized that western students of the genus would take his every word as gospel, not only because he was a Japanese botanist writing about Japanese plants, but also because it was generally assumed that he was working largely from wild plants, which he was not. Maekawa himself thought his doctorate far from the last word on hostas and, indeed, was revising it for much of the rest of his life.

Maekawa's monograph was predated by portions published from 1935 on. He took into account not only morphology, but the ecology and geographical distribution. Some authorities (e.g. Fujita) query Maekawa's division of the genus into quite so many species, and much of the material he worked with is known to have come from local gardens (*H. tokudama* and *H. undulata*). Maekawa used minute differences of flower and bract to separate species with resultant proliferation of species for which the west had, and in some cases still has, no living material. In fact Maekawa later admitted in *New Flowering Plants*, 70, pp. 3–7, that he had been too detailed in his original classification. However, he enlarged our knowledge of the genus considerably, and his Monograph 'The Genus Hosta', *Journal of the Faculty of Science*, Tokyo Univ., 1940, forms the basis of the accepted system of classification when combined with Hylander's study of European hostas (p. 50).

H. lancifolia/H. sieboldii

There was, until a few years ago, another muddle concerning the identity of *H. lancifolia*. Maekawa (1940) used Stearn's epithet *H. lancifolia* var. *thunbergii* (*thunbergiana*) but described instead the green-leafed form of *H. sieboldii*, which was named *H. spathulata* by Siebold in 1860. The Japanese themselves refer to it as *Koba-giboshi* (*H. sieboldii*), although they still do not recognize that the two species are distinct and separate. This mistake only came to light recently so that, particularly in America, the white-margined *H. sieboldii* was thought to be identical with a hypothetical *H. lancifolia albo-marginata*, which is none other than *H. sieboldii*. This accounts for Frances Williams' thinking that her *H.* 'Louisa' and others were seedlings of a variegated form of *H. lancifolia*, a plant which is in any case now known to be sterile (pp. 34 & 94). Similar problems crop up with Maekawa's botanical forms of *H. lancifolia* var. *thunbergiana* forma *kabitan*, f. *subcrocea* and f. *medio-picta*, which are now simply listed as cultivars by the American Hosta Society's Nomenclature Committee to avoid confusion, although, strictly speaking, they are botanical forms of *H. sieboldii*. Maekawa's *H. lancifolia* var. *thunbergiana* f. *albo-marginata* is *H. sieboldii* forma *typica* (which is white-margined).

According to the American Hosta Society, Japanese botanists in the 1920s were working on the theory that *H.* 'Kabitan' is a seedling of a selfed *H. sieboldii* rather than a botanical variant. However, there is still room for doubt about the origins of *H.* 'Kabitan' since its leaves, unlike those of *H.* 'Subcrocea', are narrower in relation to their length than those of *H. sieboldii* and are also more acuminate, suggesting that *H.* 'Kabitan' could be of hybrid origin with *H. sieboldii* as one parent.

Other Japanese botanists who have described some of the species include Genichi Koidzumi (1883–1953), *The Botanical Magazine*, 44, 514, 1930, who described and published the species *H. capitata* for the first time; and Jisaburo Ohwi, who published a treatment in the Smithsonian Institution's *Journal*, 1965, based mainly on Dr Maekawa's monograph. Yeiichi Araki, Maekawa's assistant based his studies on new species found in the Kinki district of Honto, the main island of Japan (1942). Araki introduces much new material although the validity of some of his names is questionable. His studies are, again, based on the work of Maekawa.

By far the most controversial of the Japanese botanists who made revisions to the genus was Noboru Fujita who made a further study of hosta classification, 'The Genus Hosta (Liliaceae) in Japan', *Acta Phytotax. Geobot.*, Vol. 27, pp. 66–96, 1976a, based on live wild specimens and dried herbarium material, reducing Maekawa's 25 species to 15 (pp. 50–1). Although Fujita's work retained Maekawa's basic taxonomic framework, his treatment leant more towards ecological differences between the species, an opinion not shared by most other botanists and taxonomists. His review was also limited by the fact that he did not include species from Korea, and also omitted some of the Japanese and Chinese species. However, he has made a significant contribution to the hosta literature.

Quite the most thorough and useful monograph on the genus though limited in its scope is Nils Hylander's 'The Genus Hosta in Swedish Gardens', *Acta Horti Bergiani*, Vol. 16, No. 11, 1954, the investigations for which he carried out at the Institute of Systematic Botany of the University of Uppsala. Prior to the monograph, Dr Hylander (1904–70) had written a preliminary account of hostas cultivated in Sweden, relying, as he admits, on the Bailey/Stearn revisions which he updated for his Swedish readers in *Lustgarden*, 1953, the Yearbook of the Society for Dendrology and Park Culture. In his monograph Dr Hylander summarizes the work of all those who have had a hand in the classification of hostas. He describes all the hostas known to him personally. Therefore, he does not cover many of the Japanese species described by Maekawa which had not then been grown in the west. His work, although thorough and in most cases accurate, is incomplete, since his range of hostas mostly came from those originally imported into Europe by Siebold. Although he made a major contribution to our knowledge of the *H. fortunei* complex, he admitted that his work was paving the way for future study, and that there were many gaps and omissions. However, I have made constant references to this valuable work.

Much of Dr Hylander's work has also been used by Dr Karel Hensen, taxonomist (now retired) at the Agricultural University of Wageningen, Holland, who has made a study of those hostas grown in the Netherlands between 1956 and 1963. He also published much useful material on the introductions of Siebold. He enlarged on his conclusions in an article in *The Plantsman*, 1985, after examining further material supplied from Japan and the United States, and herbarium specimens from Tokyo and Kyoto, Japan. He may, however, have caused unnecessary problems by placing *H. tokudama* (Maekawa) back into the *H. sieboldiana* species as *H. sieboldiana* var. *fortunei*. Dr Hensen was given much plant material from Uppsala Botanic Garden and he has corresponded with and exchanged material with hosta collectors in America.

Uppsala Botanic Garden has played an important role in the history of hostas across a period of some 200 years: in Thunberg's time in the 1790s; through to Dr Hylander's in the 1950s; and then in the 1960s, when it sent hostas to Wageningen, Holland, and thence to America.

The *montana/elata* question

The *montana/elata* question is one of the oddest of hosta muddles, and it seems that the problem begins with European systematic botanists and taxonomists trying too hard to equate hostas growing in Europe with hostas growing in Japan, for even where true species were imported originally (by Siebold for example) in many cases their places have been inadvertently usurped by hybrids cultivated in Europe; then the hybrids have mistakenly been equated with Japanese species.

The commonest hosta in Japan is called *oba-giboshi* by the Japanese, and it is generally known to gardeners and nurserymen in the west as *H. montana* (Maekawa). The name *H. montana* was published by Maekawa in his monograph in 1940 with a description.

Hylander, working in the 1940s and 1950s, presumed (incorrectly, since the comment was later retracted) that Maekawa had made the name

H. montana invalid by citing *H. bella* (Wehrhahn 1936) as a synonym for his *H. montana* var. *transiens*, the ICBN rule being that the correct name is the one first validly published. One would have supposed that Hylander would then have transferred *H. montana* to *H. bella* but this he did not do. Instead, he proposed the name *H. elata* which he duly published together with a description in English. Since the name *H. bella* was validly published and there might be some doubt as to whether *H. elata* is valid for the taxon proposed, one would suppose the correct scientific name for *oba-giboshi* to be *H. bella* (Wehrhahn 1936).

But it is not enough in synonymy merely for a name to have been validly published: there must also be substantial similarities between the plants described. In the case of *H. montana/bella/elata* this is questionable.

Hylander, a European botanist working with European garden material, equates his *H. elata* with the *H. montana* of Maekawa, though he had never seen the true *H. montana*, and bases his equation on parts – not the totality – of Maekawa's description and admits 'I feel … uncertain … whether … *H. montana* s.str. are the same species as my *H. elata* …'

If *H. montana* (Maekawa 1940) is not the same plant as *H. elata* (Hylander 1954), what, one wonders, is *H. bella* (Wehrhahn 1936)? Maekawa, who created the synonymy, apparently never saw *H. bella*. But then Hylander, who makes *H. bella* a synonym of *H. fortunei* var. *obscura*, never saw the true *H. bella* 'since I have not seen any authentic specimen of that form', (Hyl. 1954).

In view of such vagaries one is forced to compare type specimens – which none of the botanists did. One then finds that *H. bella*, now sadly a lost cultivar, has 8–10 pairs of veins and *H. montana* var. *transiens*, with which Maekawa says it is synonymous, has 14–15: so they are plainly not the same plant. *H. elata* in Hylander's diagnosis does have 8–10 veins, but then Hylander makes *H. elata* a synonym of *H. fortunei* var. *gigantea*, though the specimens differ too greatly for them to be the same plant. Nor does Hylander's description of *H. elata* have sufficient points of similarity with Maekawa's diagnosis for them to be the same plant.

So in spite of the plethora of synonyms, and incredible though it may seem, the commonest hosta in Japan, *oba-giboshi*, still has no valid scientific name, because Maekawa's name is invalid.

Hosta 'specialists'

Mrs Frances R. Williams

Hostas were first imported into America from Japan by Thomas Hogg (*q.v.*) and have been grown there ever since. They were much used in Victorian and Edwardian shrubberies and cemeteries but were not considered fashionable or important at that time. It was not until the 1950s and 1960s that the landscape architects who design small city gardens and shaded courtyards began to see the merits of hostas, although one of their number, Mrs Frances Williams, had been amassing a collection since the 1930s. It is mainly due to her enthusiasm that they are now one of the most popular perennial plants in the United States.

Frances Ropes Williams graduated from the Massachusetts Institute of Technology with a degree in landscape architecture in 1904, a time when few women took higher education seriously. She worked for two years with Warren P. Manning, landscape architect. Her studies at MIT gave her a basic knowledge of the way plants grew, their shapes, colours and relation to sun and shade.

She married Stillman P. Williams and then devoted the next 30 years to raising her family. As soon as time permitted she began to take an interest in her own shaded garden ($\frac{1}{3}$ acre [0.13 hectares]), and in hostas in particular. Some of the hostas she grew came from friends in Salem, others she obtained from local nurseries. Taking short drives from her home, she visited the nurseries of Gray & Cole at Ward Hill, Massachusetts. Gray & Cole were becoming interested in plantain lilies, as hostas were once known, as was Mr Charles Stockman of Newburyport. Both gave her plants. She began corresponding with botanists, taxonomists and leading horticulturalists in several countries and as she was willing to share her own hostas with others, she naturally became the recipient of many plants and seeds from different parts of the world. She meticulously labelled and catalogued all these plants and also kept accurate records of the many hybrids she raised. These records were donated to the Minnesota Arboretum Library (p. 161).

Her most important contribution to the genus is the hosta known as H. 'Frances Williams', which she noticed in a nursery bed at the Bristol Nurseries, Bristol, Connecticut on 17 September 1936. As she records it, 'In some fine grey-leafed hostas there was one plant with a yellow edge'. Two workmen dug it up for her on the spot. It was growing in a long row of grey-green *H. sieboldianas*. She bought the plant under the name 'Glauca gold' and called it *H. sieboldiana* 'Yellow Edge'. It is now presumed that the row of grey-green plants in question was actually the form 'Elegans', of which this yellow-edged plant appears to be a sport rather than a seedling.

Mrs Williams returned part of the plant to Bristol Nurseries. She also gave pieces of it to many people, including Gray & Cole and Mrs Thomas Nesmith of Fairmount Gardens, Lowell, Mass. Mrs Williams stated that H.A. Zager, a nurseryman of Des Moines, Iowa, who also received this hosta from her, 'himself put the Latin name on the yellow-edged *H. sieboldiana.*' Zager's name was *Aureus marginata*.

The name 'Frances Williams' did not exist until 1963. Mrs Williams had been corresponding with Mr George W. Robinson, superintendent of Oxford Botanic Garden, and with Mr Graham Stuart Thomas, to whom she sent copious notes on her observations. In these notes she mentions finding this hosta and in 1959 she sent a piece to Oxford Botanic Garden. It is understood that Mr Robinson forgot the name under which Mrs

Williams sent him the hosta and, in a lecture on variegated plants presented to the Royal Horticultural Society, referred to the plant as *Hosta* 'Frances Williams'. Robinson used the name again in the published version of his lecture which appeared in *The Garden*, (the Journal of the Royal Horticultural Society) in February 1983, and so the name was published and firmly stuck. It is now the preferred name in the USA, and has, at last, recently been registered. It is the most popular hosta ever raised.

H. 'Frances Williams' is not, however, the first yellow-edged, glaucous-leafed hosta on record. Mrs Gertrude Wister in her notes on *H.* 'Frances Williams' in the 1969 issue of the *American Hosta Society's Bulletin* quotes Mrs Williams as saying, 'Looking back over some old notes I find a statement by Charles Stockman of Newburyport, that he had seen a yellow-edged hosta in Germany. On 23 September 1935, Mr Stockman said he had seen these in Munich's Botanic Garden'. These plants could have been the hostas that Karl Foerster, the East German nurseryman who was responsible for raising the gold-leafed *H.* 'Semperaurea', listed in his catalogue in the 1930s as *H. sieboldiana fortunei* 'Aureo-marginato' or *Blaue Gelbrandfunkie* (= blue yellow-margin funkia) which could have been the same as *H.* 'Frances Williams' but could have been what we now know as *H. tokudama* forma *aureo nebulosa*, often listed in Europe at that time as *H. sieboldiana* var. *fortunei* 'Variegata'. (It could also have been *H.t.* forma *flavo circinalis* which can arise as a sport from *H.t.* f. *aureo nebulosa*).

William Robinson in his *The English Flower Garden*, John Murray, 1883, mentioned a *Funkia sieboldii* with a gold edge and, as discussed earlier, this was then the often-used epithet for *H. sieboldiana*. This gold-edged hosta could, again, have equally well been a gold-edged form of *H. tokudama*.

Mrs Williams raised many other hostas from seed, some of it Japanese. She named many of these seedlings, often after members of her family and friends. Some are still listed by nurseries but most have been superseded. *H.* 'Sweet Susan' is one of the most interesting, as it was the first hosta raised with fragrant, mauve flowers. The others which are still thought to be good garden plants are considered in the chapter on cultivars.

Eric B. Smith (1917–1986)

Eric Smith, a vice-president of the British Hosta & Hemerocallis Society, is the only British hosta hybridizer to have achieved lasting fame. He is best known for his inter-specific cross *H.* × *tardiana*, a name which although universally known and used is in fact illegitimate. The Nomenclature Committee of the British Hosta & Hemerocallis Society has now agreed that this interspecific cross should be known as the Tardiana group, a name with conforms with ICBN rules.

Eric Smith was employed by Hilliers of Winchester as a propagator where it was discovered that he had a talent for raising exceptionally good hybrids. At Hilliers he raised his first well-known hosta, *H.* 'Buckshaw Blue', though it was not named until he established with Jim Archibald The Plantsmen nursery, whence it was first marketed. It is one of the few hostas to have won an Award of Merit after trial at Wisley.

Although Eric Smith's major contribution to the genus hosta is his Tardiana group, he was amongst the forerunners of hosta breeders who saw the possibilities of gold-leafed forms. His gold-leaf strain, the GL series, is the result of experimental crosses using *H. fortunei* var. *albopicta* forma *aurea* with a near-white-flowered form of *H. sieboldiana* var. *elegans*. His ambition was to raise a hosta of good substance and flowers which retained gold-coloured leaves right through the summer. *H.* 'Gold Haze' and 'Granary Gold' are among the best of these.

Working with a different cross, *H. sieboldiana* var. *elegans*, (again the white-flowered form), and *H.* 'Kabitan', he raised seedlings of a harsher golden hue, widely spread venation and thin texture. Of these *H.* 'Hadspen Samphire' and *H.* 'Goldsmith' were very near to his aim, but the golden leaves of these hybrids do become a dull chartreuse-green towards the end of summer. These latter hybrids scorch in sun and must be grown in shade.

Both these lines are now largely superseded as breeding for hostas with gold leaves is being done with golden sports of *H.* 'Frances Williams' and *H. tokudama* forma *aureo nebulosa*, but much valuable information has been gleaned from Eric Smith's work in this field.

One of the few hostas raised at Buckshaw Gardens was *H.* 'Snowden', a large, greyish-green-leafed seedling, derived from *H. sieboldiana* and an *H. fortunei*. Although a collector's plant and a majestic one at that, *H.* 'Snowden' has never quite attracted the acclaim that the similar-sized *H.* 'Krossa Regal' has (Fig. 116).

Undoubtedly the Tardiana group represents Eric Smith's finest achievement. He raised the first of these small, glaucous-blue hostas whilst he was still at Hilliers and he continued to raise new generations during his ten years at The Plantsmen and later as head gardener at Hadspen House. His success with this cross was a lucky accident. In the early autumn of 1961 he found a late flower on a young plant of *H. sieboldiana* 'Elegans Alba' (normally a June or July flowerer which occasionally re-blooms) and, on the spur of the moment, put some of the pollen on a fairly early flower of *H. tardiflora* which usually flowers in late September and October and sometimes even into November, hence its name. He had to cut the spike before the seed was ripe and finish it off in a solution of sugar and water. He had been planning to try to raise small glaucous-leafed hybrids for some time, as *H. sieboldiana*, good garden plant that it is, is too large for many smaller gardens, and he could see the aesthetic and commercial potential for dwarf and small, neat, blue-leafed hostas.

From the first cross he raised about 30 seedlings, of which 14 were glaucous. The best of these are TF1x1 'Dorset Charm', TF1x4 'Dorset Flair' and TF1x7 'Halcyon'. He did not bother to keep the other clones separate, apart from TF1x5 'Happiness', and he sent these three to the hosta trials at Wisley. *H.* 'Halcyon' gained an Award of Merit.

He attempted on many occasions to repeat his cross, and to self *H. tardiflora*, and used the pollen of any other species (including the *H.* x *tardiana* F1s) which happened to be flowering at the right time, but never obtained a single seed. The F1 x *tardianas* were easy to grow and made vigorous root systems; they produced abundant seed which germinated well.

He next crossed TF1x1 with TF1x4 and got even better results. As can be expected from a cross between two such different hostas, even a generation back, the TF2s proved to be very variable in size, colour and shape. Both glaucous and green-leafed forms occurred. Most of the latter were of medium size (about the same as that of the TF1s) and were not distinctive enough to have a use other than as ground-cover plants. The glaucous F2 seedlings, however, produced some attractive forms, including a few neat, smallish ones with dimpled or puckered leaves (TF2x8 'Blue Dimples'), which he selected for vegetative propagation. Most were similar in size and shape to *H. tardiflora* ('Hadspen Heron' TF2x10 and 'Hadspen Hawk' TF2x20), but two others, TF2x2 'Blue Moon' and TF2x4 'Dorset Blue' have small rounded leaves resembling the original pollen parent, *H. sieboldiana* 'Elegans Alba'. *H.* 'Hadspen Blue' (TF2x7) is probably bluer than *H.* 'Halcyon', which is supposed to be one of the best of all.

Most of these small hostas produce flower scapes not more than 8 in (20.5 cm) high, and some barely overtop the leaf mound. The best flower, a good purple with a dark stripe down the middle of the petal is on TF2x3 'Harmony'. Eric Smith also used TF1x7 as a parent in order to produce more of the smaller types. Most of these seedlings are sterile or nearly so, but he did manage to raise a few seedlings from one of them (TF2x16), of which only TF3x1 'Blue Blush' is a good dwarf, narrow-leafed form. It produced just two pods but no seed. Eric Smith thought that the plants seemed to be getting smaller with each successive generation. TF3x1 has leaves and flower spikes only about 3 in (7.5 cm) long. It flowered for the first time in 1975. In spite of the fact that the *H. sieboldiana* 'Elegans' used in the original cross was white-flowered, only two of the resultant seedlings have pure white flowers and both are slow to increase and also sterile (TF2x14 ['Osprey'] and TF2x30 ['Brother Ronald']).

Eric Smith corresponded with and exchanged hostas with Alex Summers, then President of the American Hosta Society. Mr Summers visited Eric Smith at Hadspen House and chose what he considered the 12 best of the Tardiana group and took them back to the United States. He and Eric Smith had named them together, some with the prefix 'Dorset' and some with the prefix 'Hadspen', and yet others had the prefix 'Blue'. One or two were registered with the International Hosta Registration Authority at Minnesota but for reasons unknown others were not. The British Hosta & Hemerocallis Society is currently undertaking the registration of all the

Fig. 6 *H. lancifolia*
Description on p. 94

Tardiana group which have TF numbers in order to safeguard the original names. The Society is also naming those still known only under numbers growing in the National Hosta Reference Collections and others known to be in existence from Eric Smith's own notes. The names chosen have either a 'Sherborne' prefix or the name of a bird: one or two have geographical names, for example 'Purbeck Ridge' (originally named by Julie Morss). Forms with a mottled central cream variegation now exist of which *H.* 'Goldbrook Glimmer' is one, and there is a cross between TF2x10 ('Hadspen Heron') and *H.* 'Kabitan' which Eric Smith named H. 'Golden Oriole'. It is similar to *H.* 'Hadspen Samphire' in general appearance and size, having dark red-spotted petioles, undulating lance-shaped leaves which are shiny above and below, and it flowers in late summer. It is not available commercially.

The arrival of the Tardiana group in the States during the 1970s was hailed as a breakthrough in hosta hybridizing and the plants were seized on by nurserymen and collectors alike.

Eric Smith gave away a considerable amount of F1 seed and unnamed plants, so it is likely that there may be good examples of the Tardiana group growing in various parts of the world. The problem is that they all have seemingly similar characteristics but will be given different names and thus add to identification problems unless named plants are properly registered with the correct authority. It should be possible in the near future to ensure such errors do not happen once all the standard specimens are available for inspection in the National Reference Collections (p. 156).

It seems more than likely, the author having had access to Eric Smith's useful but brief notes and to his slides (these notes and slides are jointly held by the British Hosta & Hemerocallis Society and the main National Reference Collection) that some of the hybrids have been lost to cultivation: one of the named GL Series *H.* 'Dorset Cream'; an unusual cross between *H. fortunei* 'Albo marginata' and *H. f.* 'Albo picta' which he named *H.* 'Iceland' on account of its very white leaf-centre; and *H.* 'Sea Wave' and *H.* 'Yellow River', two of his early hybrids which he listed as growing in his Southampton garden, are missing, as is *H.* 'Goldsmith'.

3.

THE HOSTA SPECIES

A short introduction

In 1987 a new system of classification for the petaloid monocotyledons was introduced by the Royal Botanic Garden, Kew. This new classification is largely based on the system put forward by R.M.T. Dahlgren, H.T. Clifford and P.F. Yeo in their book *The Family of the Monocotyledons* Springer-Verlag, Berlin & Heidelberg, 1985. They placed the genus *Hosta* in *Funkiaceae* of which it is the sole member. The family name *Funkiaceae* is based on *Funkia* Sprengel (1817) but is illegitimate (p. 40). Brian Mathew has therefore proposed the family name *Hostaceae* B. Mathew (1987), based on *Hosta* Trattinick. Hostas formerly belonged in the family *Liliaceae*. The genus *Hosta* itself remained unchanged.

However, most authors have found it necessary to further classify hostas at a level between genus and species which they have variously called sections or subgenera or both. The purpose of such further classifications, as indeed of all systematic botany, is to try to arrive at natural groups, evolutionarily, ecologically and geographically. The first such attempt was made by Sir Edward Salisbury (1812), (p. 38). He split the genus into two new genera, one of which – the genus *Niobe* – has been retained by all subsequent authors as a subgenus or section. It contains the single Chinese species, *H. plantaginea*. The rest of the genus has been addressed variously by different authors.

Salisbury also created the genus *Bryocles*. Bailey (1931) took matters a little further forward when he proposed splitting the second of Salisbury's new genera into two sections, *Bryocles* and *Alcyone*, with *H. ventricosa* as the type plant of the section *Bryocles*, as it was of Salisbury's genus of the same name. Maekawa went a step further (1937) when he sub-divided *Bryocles* into section *Bryocles* and sub-section *Eubryocles*, citing *H. ventricosa* as the type plant. In 1938 he further proposed the subgenus *Giboshi* for the remainder of the species, and this sub-section equates well with Bailey's *Alcyone*. Thus if we accept Maekawa's placement of *Giboshi* it is equally valid to accept his *Bryocles* (1937) on grounds of priority. There is thus a substantial measure of agreement between Bailey and Maekawa at supra-specific level, Bailey's subgeneric groupings being *Niobe*, *Bryocles* and *Alcyone*: Maekawa's *Niobe*, *Bryocles* and *Giboshi*. However, Maekawa in his major work on the genus (1940) reduced the number of subgenera to two, an arrangement which now seems artificial and unnatural, although he did revert to three subgenera in all later publications. Unfortunately this is not generally known in the western world since these later works have not appeared in an English version. Most western students of the genus have therefore relied solely on his 1940 placements.

But there are still problems. Maekawa cited no type for the subgenus *Giboshi*. Hylander (1954) who worked only with hostas of European garden origin and those introduced from

Japan by Siebold, proposed *H. sieboldiana* as the type plant of *Giboshi*, and this was generally accepted. It turns out that Hylander based his proposals on Fujita's distribution maps of *H. sieboldiana* var. *sieboldiana*, but *H. sieboldiana* is a name of European origin and misunderstandings can arise when it is applied to wild populations in Japan. Fujita applies the Japanese academic name *Oba-Giboshi* to his *H. sieboldiana*, and from this it is plain that the plant he intends is in fact *H. montana*. It certainly makes far better sense to accept *H. montana* as the type plant of the *Giboshi* subgenus since *H. montana* is as common a hosta in the wild in Japan as *H. sieboldiana* is rare.

It further turns out that a division of the genus into three subgenera – based mainly on morphological criteria – makes sound geographical sense since the subgenus *Niobe* contains a single distinct Chinese species, *H. plantaginea*, subgenus *Bryocles* contains the north Korean and Chinese species, and subgenus *Giboshi* the Japanese species. This leaves us with two anomalous hostas, *H. jonesii* and *H. yingerii*, both very variable-leafed, spider-flowered Korean species new to science. It is not yet at all clear where these species fit. Here, they are placed provisionally in a section of their own, *Arachnanthae*, in the subgenus *Bryocles*.

According to whether one is a 'lumper' or a 'splitter' there are between 20 and 50 species of hosta. New hostas are still being found in the wild, mainly in Korea, although most of these are in the hands of collectors, mainly nurserymen, who give them fancy or cultivar names or even pseudo-scientific Latin names. Some are perhaps species and await proper diagnosis. It is, therefore, impossible to be exact as to the number of species. Hostas hydridize freely with each other and further blur the distinctions between the species which is another reason for uncertainty over numbers.

Below is a summary of the subgenera and sections to which the various species are currently consigned, but no taxonomic arrangement is ever final.

Subgenus *Niobe*

Based on Salisbury's original genus and universally recognized as distinct from all other hostas. The Japanese call it *Maruba-tamano-kanzashi* distinguishing it from other hostas which are called *Giboshi*. It is unique in its long-tubed flowers. Found only in China.

H. plantaginea is the sole species in this subgenus.

Subgenus *Bryocles*

(a) SECTION EUBRYOCLES

Originated in the Lu-Shan mountains in China and in Korea with escapes to north Nagasaki and perhaps Kyushu. The bracts are navicular, grooved, thin, dry and membranaceous.

H. ventricosa is the sole species in this section.

(b) SECTION LAMELLATAE

(i) Sub-section Capitatae.
Scapes are hollow. The flowers, while in bud, are gathered into a dense, ball-like head at the top of the scape. Bracts are densely imbricated and purple in colour. The distribution for this section is Korea and China with escapes to Nagasaki Prefecture.

H. capitata, H. nakaiana.

(ii) Sub-section Spicatae.
Scapes are solid with thin grooves. Leaf blades are somewhat thin and papery. Bracts navicular. The distribution is the same as for Sub-section Capitatae.

H. minor, H. venusta.

(c) SECTION ARACHNANTHAE

Species provisionally assigned to this Section have recently been found in Korea. They are anomolous in possessing spidery flowers and glossy leaves. The distribution for this coastal section is the Taehuksan district.

H. jonesii, H. yingerii.

Subgenus *Giboshi*

(a) SECTION HELIPTEROIDES
So called from a similarity in the conformation of the bracts with those of the genus *Helipterum* (Compositae) in which the flowerheads are surrounded by numerous, large, persistent overlapping bracts. Found mainly in the lowlands of Honshu but distributed as far westwards as Kyushu and northwards to Hokkaido. The flowers of the species in this section are near-white, the scapes round and solid. All, except *H. elata* and *H. crispula* show varying degrees of mealiness or pruinosity, and the veins on the underside of the leaf blades are very prominent.

H. montana, H. crassifolia, H. elata, H. fluctuans, H. crispula, H. nigrescens and *H. sieboldiana* as well as *H. tokudama* and *H. fortunei.*

(b) SECTION RHYNCHOPHORAE
From the Greek, *rhyncho* ('provided with a snout or beak'), and *phora* ('a carrier'). The name reflects a fancied resemblance to the beaks of the Japanese Kutari cranes. The beak-shaped bud is enveloped by sterile bracts. The distribution band is narrow, i.e. from Yakushima Island to Hyuga, Tosa and Iyo. This section is distinguished by an abbreviated raceme and straight, flat bracts, causing the raceme to appear bird-like (most pronounced in *H. kikutii* var. *caput-avis*). The scapes are solid and leaning.

H. kikutii, H. pycnophylla, H. shikokiana.

(c) SECTION INTERMEDIAE
Believed to have evolved from the Helipteroides

section. Grows throughout Hokkaido region. Scapes are solid, bracts densely imbricated (overlapped), green, navicular. Corolla is thin-textured, tube round and the scape tends to be leafy.

H. kiyosumiensis, H. densa, H. pachyscapa.

(d) SECTION STOLONIFERAE
This section has an affinity with the Tardanthae section. It occurs in Korea and is distributed northwards to Kirin Province, Manchuria. Noted for its very long stolons which reach far out underground. Leaf blades are somewhat stiff, held erect and dark green in colour, veins being scarcely impressed on the upper surface. Scapes are solid. Bracts are grooved and navicular, greenish-white and remain the same shape before and during flowering. Corolla tubes are grooved. A horticultural substitute in the Korean Peninsular for section Nipponosta.

H. clausa.

(e) SECTION PICNOLEPIS
Pynco, is Greek for 'close', 'dense', 'compact'. *Lepis* is a 'scale': therefore 'having close (dense) scales on the bracts'. This section extends from Tochigi Prefecture (Shimotsuke Province) Central Honshu through the mountainous region of West Kanto, extending to Kyushu in the west and Hachijo Island in the south. Plants of this section are often found growing on rocks or in crevices. Scapes are solid, the narrow angular corolla is thin-textured with a cylindrical, ungrooved tube. The bracts are membranaceous, whitish or purple in colour and navicular. The leaf blades are stiff-textured, the veins scarcely impressed on the upper surfaces. They are late to flower.

H. longipes, H. rupifraga, H. tardiflora, H. sparsa, H. hypoleuca, H. tortifrons, H. okomotoi, H. takiensis.

(f) SECTION TARDANTHAE

Named from the Greek *tardus* ('late'), *anthos* ('flower'). This section begins in the Kinki district and, passing through Shikoku to Kyushu, has isolated representatives as far away as the Kiangsu and Chekiang Provinces of mid-China. The scapes are very rigid and round, the bracts navicular, thin, dry and membranaceous. The narrow corolla is grooved. All species in this section are late to flower.

H. tardiva, H. okamii, H. gracillima, H. pulchella, H. tibai, H. takahashii, H. tsushimensis, H. cathayana.

(g) SECTION NIPPONOSTA

Distributed over Kyushu, Shikoku and Honshu. Some plants occur northwards to Hokkaido, South Kuriles, Sakhalin and the Maritime Provinces. Nearly all tend to be small plants. Leaf blades are narrow and lance-shaped, and the scapes are round and solid. The bracts are grooved and navicular, greenish-white and remain the same before and during anthesis. Corolla tube is grooved.

(i) Sub-section Nipponosta-Vera (distinguished by different form of perianth):

H. longissima, H. helonioides, H. rohdeifolia, H. ibukiensis, H. lancifolia, H. clavata, H. alismifolia, H. calliantha, H. campanulata, H. sieboldii.

(ii) Sub-section Brachypodae, *Brachy* ('short'), *podos* ('foot'). *Brachy-podus* = 'short-stalked'. Occurs in the highland region of central Japan. Distinguished by green, navicular, densely imbricated bracts.

H. decorata, H. rectifolia, H. opipara, H. atropurpurea.

(h) SECTION FOLIOSAE

Plants with large, leafy bracts on the scapes, the bracts imbricated, variegated white/green or green/white, navicular. Scapes solid. Corolla thick, lilac in colour. It is an anomaly that this section is based on garden plants, when one would expect a botanical division like this to be based on wild plants. Perhaps the answer is that the plants in this section are so distinct as to warrant this treatment, regardless of their origins. They have been grown in gardens in Japan since the remotest times. Recent, but unconfirmed reports that hostas of this section have been found growing wild on Mt Kurami turn out to be wrong. The original habitat may turn out to be Korea and the plants have migrated over the centuries to Nagasaki Prefecture

H. undulata.

Subgenus *Niobe* (Salis.) Maekawa

H. plantaginea (Lamarck 1789)
Ascherson 1863

Hemerocallis japonica Redouté 1802
Hemerocallis alba Andrews 1801
Funkia subcordata Sprengel 1825
Hosta 'Thomas Hogg' (misapplied)
August Lily of cottage gardens
Chinese name: *Yu-san*
Japanese name: *Maruba-tamano-Kanzashi*

Called 'plantaginea' from the fancied similarity of the leaves to those of plants of the plantain family *Plantaginaceae*, which are longitudinally ribbed. It is quite distinct in general appearance from other hostas, a fact recognized by the Japanese who call it *Tamano-giboshi-kanzashi* as opposed to *Giboshi* which they use for all other hostas.

SUMMARY OF THE SPECIES IN THEIR SECTIONS

Subgenus	Section & Sub-section	Species
1. *Niobe*		*H. plantaginea*
2. *Bryocles*	(a) **Eubryocles**	*H. ventricosa*
	(b) **Lamellatae**	
	(i) **Capitatae**	*H. capitata, H. nakaiana*
	(ii) **Spicatae**	*H. minor, H. venusta*
	(c) **Arachnanthae**	*H. jonesii, H. yingerii*
3. *Giboshi*	(a) **Helipteroides**	*H. montana, H. crassifolia, H. elata, H. fluctuans, H. crispula, H. nigrescens, H. sieboldiana, H. tokudama, H. fortunei*
	(b) **Rhynchophorae**	*H. kikutii, H. shikokiana, H. pycnophylla*
	(c) **Intermediae**	*H. kiyosumiensis, H. densa, H. pachyscapa*
	(d) **Stoloniferae**	*H. clausa*
	(e) **Picnolepis**	*H. longipes, H. rupifraga, H. tardiflora, H. sparsa, H. hypoleuca, H. tortifrons H. okomotoi, H. takiensis*
	(f) **Tardanthae**	*H. tardiva, H. okamii, H. gracillima, H. pulchella, H. tibai, H. takahashii, H. tsushimensis, H. cathayana*
	(g) **Nipponosta**	
	(i) **Nipponosta-vera**	*H. longissima, H. helonioides, H. rohdeifolia, H. ibukiensis, H. lancifolia, H. clavata, H. alismifolia, H. calliantha, H. campanulata, H. sieboldii*
	(ii) **Brachypodae**	*H. decorata, H. rectifolia, H. opipara, H. atropurpurea*
	(h) **Foliosae**	*H. undulata*

The typical *H. plantaginea* comes from China and was the first hosta to be grown in Europe. Its historical importance is discussed more fully on p. 36. Certainly by the end of the 18th century it was in cultivation in Britain though it was grown originally as a stove plant and was only tried out of doors much later.

H. plantaginea is an exception to the general rule that hostas need to be grown in at least some shade and grows best, as it does at the Savill Garden, Windsor Great Park, at the foot of a south-facing wall. Unless it is given such a position, or grown in a tub against a wall, it simply will not flower regularly north of London nor in the northern parts of Europe. It excels itself in the south of France where the climate is too hot for most hostas. Although very fertile and the parent of many hybrids, seed is only formed infrequently in the UK, with only one or two pods per stalk. It obviously prefers a hotter climate. This is because by the time it has flowered – late August to the end of September and, occasionally into October – the seed does not have time to ripen before the cold weather and frosts arrive. Although it has glistening, fresh lime-green leaves, it is for the exquisite fragrance of its night-opening flowers that it has been such a favourite for so many years. It is most often grown as a cottage-garden plant, mingling with scented day-lilies, roses and phlox, rather than in a shade garden or more formal hosta planting.

*

A robust species forming an open clump. Rhizome thick, roots compact, round, long, white. Petiole 10 in (25 cm) long, deeply channelled, winged with overlapping edges on upper part. Blade basically ovate, but cordate and decurrent at the base, apex shortly acuminate; margin very shallowly undulate; up to 11 × 9½ in (28 × 24 cm) in America but more typically 9 × 6½ in (23 × 16 cm) in Britain; glossy, yellow-green above, glossy underneath with 6–9 pairs of raised, widely spaced, well-marked veins, devoid of mealiness. Scape stout, solid, bright green, non-mealy, up to 25½ in (65 cm), rising well above leaf mound. Foliaceous bracts few. Flower bracts greenish-white. Flowers night-opening, on short, dense racemes, tubular, 2½ in (6 cm) long, glistening, waxy-white, fragrant. Anthers small, light yellow. Capsules large, cylindrical, 2½ (6 cm) long, rarely produced, but when produced very fertile. Flowers late summer to early autumn.

Illustr.: Acta Horti Bergiani: 16: Pl. 24 (1954); *Hosta: The Flowering Foliage Plant,* Fig. 14.

Var. *grandiflora* (Lemaire 1846) equals Var. *japonica* (Kikuchi & Maekawa 1940) and, incidentally, has priority. Has slightly larger, longer, flowers and narrower leaf blades which have a lax habit, giving a sprawling untidy appearance to the leaf mound. (Award of Merit 1948).

Forma *stenantha* (Maekawa 1940) has shorter, narrower corolla tubes. Found by Dr Kikuchi in Paoting, Hopeh, China.

Forma 'Aphrodite' (Maekawa 1940) is a so-called double and has flowers with petaloid stamens. It comes from northern China and was only recently introduced to Europe and America from Beijing.

H. 'Chelsea Ore' (Compton) found in the mid-1980s at the Chelsea Physic Garden, London is a gold-leafed form of var. *grandiflora* with a narrow, green margin.

Subgenus *Bryocles*   Maekawa (1937)

SECTION EUBRYOCLES

H. ventricosa	Stearn (1931)
Hemerocallis caerulea	Andrews (1797)
Funkia caerulea	Andrews (1827)
Funkia ovata	Sprengel (1825)
Funkia latifolia	Miquel (1869)
	Andrews (1891)

Saussurea coerulea Nash (1911)
Niobe coerulea Japanese name:
  Murasaki-giboshi

One of the best flowering plants in the genus and one of the best of the green-leafed hostas, being of good poise and shape and having distinct and attractively urn-shaped, deep purple flowers, hence Salisbury's name *Bryocles ventricosa* on account of the suddenly swollen (ventricose) flower tube. It is sometimes mistaken for *H. undulata* var. *erromena* and *H. fortunei* var. *stenantha*, but it is a deeper green and the veins are spaced further apart, although the glossy leaf surfaces often appear similar on a superficial examination.

One of the few hosta species which is indigenous to China and not Japan and, historically, the second hosta to be introduced to Britain and Europe. Its history and introduction is fully discussed on p. 36. It was given the Japanese academic name by Dr Kikuchi in 1937.

H. ventricosa prefers dappled shade but will grow in more sun in British gardens if there is sufficient moisture in the ground, but in its variegated forms it tends to scorch at the margins if the sunlight is too intense.

H. ventricosa and its variegated forms are the only hostas known to be apomictic. Apomixis is a modification of the normal reproductive process by which the plant produces viable seed without cross-pollination. This is, therefore, a form of natural clonal propagation.

It means that *H. ventricosa* breeds true (though conversely it also means it cannot produce a cross as a pod parent). As a pollen parent it will produce hybrids as the pollen is excellent. Cultivated plants are tetraploid (2n = 120).

*

Vigorous, medium-sized to large clump-forming hosta. Rhizome compact, non-fusiform, compactly clustered with cylindrical, long white roots. Petiole up to 16 in (40 cm), broad, and broadly grooved, green, pale rose-purple dots near base. Blade stiff, widely ovate or cordate, apex sharply acuminate, margin slightly undulate, base blunt, decurrent; up to $9\frac{1}{2} \times 7$ in (24 × 18 cm) wide; thin substance, glossy dark green above, shiny mid- to dark green below with 7–9 pairs of widely spaced veins and conspicuous, small transverse veins. Scape round, rigid, erect, base suffused red; up to 36 in (103 cm), one foliaceous bract towards centre. Flower bracts broad, green with white base, thin and withering by anthesis. Raceme long, lax, bearing 20–30 urn-shaped, deep bluish-purple flowers. Anthers purple-spotted. Fruit long and thick. Seeds large. Flowers late summer.
Habitat: Kiangsu, Kwantung, China.
Illustr.: Curtis's Bot. Mag., 894, 1806; *Gentes Herbarum* 2, 128, 1930; *Hosta: The Flowering Foliage Plant*, Fig. 2.

H.v. 'Aureo-maculata'. Mentioned by Miquel as long ago as 1869 under the name *Funkia latifolia* lusus *aureo-maculata*. This has always been a rare but much sought-after hosta. Smaller than the type, leaf blade bright yellow in the centre, surrounded by an irregular dark green margin. The yellow variegation fades gradually during the summer but never quite disappears. Flowers identical with type.

H.v. 'Variegata' (known in America as *H.v.* 'Aureo-marginata'), has creamy-white irregular margins and is still considered one of the best of the variegated hostas. Flowers identical with type. Discovered by Alan Bloom, VMH, in a nursery bed of *H. ventricosa* in a Dutch nursery before 1968 and introduced by Blooms of Bressingham Gardens.

SECTION LAMELLATAE

Sub-section Capitatae

H. capitata Koidzumi (1930)
  Nakai
H. caerulea var. *capitata* Koidzumi (1916)

H. nakaiana   Maekawa (1940)
                       Japanese name:
                         Iya-giboshi

H. nakaiana   (Maekawa 1935)
                         Maekawa 1937
                       Japanese name:
                         Kanzashi-giboshi

The type plant of the section and the largest plant in it. It has distinctly scalloped margins. The flower buds exhibit the same characteristics as those of the only other member of the section (*H. nakaiana*). It is not well-known in Britain where it is regarded as more of a collector's item than as a good garden plant, but it can be a welcome and interesting addition to a shaded part of a rock garden or peat bed. In the wild it inhabits limestone outcrops, but in spite of this it is one of the least drought-tolerant hostas.

*

Small to medium-sized hosta forming clumps. Leaves sprout early in the season. Rhizome thick with stout radial fibres. Petiole slender, 4 in (10 cm) long, deeply grooved, spotted purple. Blade cordate-ovate; up to 5 in (13 cm) long × 3 in (8 cm) wide; margins conspicuously undulate; almost ruffled; apex abruptly acuminate; of thin substance, dark olive-green, opaque on upper surface, lustrous beneath; with 7–9 pairs of veins, impressed above, small papillae clearly evident on underside of veins. Scapes purple-dotted, ridged to the touch, erect-solid, about twice as long as the blade and can reach up to 16 in (40 cm) tall with 3 navicular foliaceous bracts just below the inflorescence. Flower bracts elliptic, intense green before anthesis, then thin and withered. Raceme short and compact, congested in bud and nearly capitate. Flowers tubular to funnel-shaped, dark purple, narrow section of tube whitish. Anthers small, purple. Fruit large and heavy. Flowers mid-summer. Remontant when mature and growing in a warm climate.

Habitat: Korea and also Chugoku, Shikoku and Kyushu in Japan.

Illustr.: None available.

Named in honour of Professor T. Nakai (who collected it) by Maekawa who was both his pupil and son-in-law. The Japanese name *Kanzashi-giboshi* means 'ornamental hair piece hosta', no doubt referring to the flowerhead described by Maekawa as 'large lilac flowers which make a head bisect by closely imbricated purple bracts'. This description applies equally well to *H. capitata*.

An elegant species with scapes far overtopping the leaf mound. All the flowers in bud are clustered in a bunch at the top of the scape, instead of being spread out along the upper section of the scape as is more usual in hostas. At the time of writing the only specimens of *H. nakaiana* growing in British gardens are seed-raised plants.

H. nakaiana increases rapidly and is the parent of many good cultivars, some of which are described in Chapter 4, including a gold-leafed cultivar 'Birchwood Parky's Gold' which has been distributed in Britain erroneously as *H. nakaiana* 'Golden'. (Fig. 7) *H.* 'Birchwood Parky's Gold' must be a hybrid and not a form or variant of the species since it does not have the characteristic ridged scapes. *H. nakaiana* is a graceful garden plant which deserves to be better known.

*

Plant small, clump-forming. Rhizome thick, with thick nodes, bearing a ring of bristles left by the decaying petioles. Petiole slender, 4 in (10 cm) long, deeply grooved, narrowly winged, purple-dotted near the base. Blade cordate-ovate, cordately truncate at the base, apex sharply recurved, margin undulate, 2½ in (6 cm) long × 1½ in (3.5 cm) wide; slightly leathery texture, of thinnish substance, opaque on upper surface, pale and shiny below; membranaceous and papery texture, with 5–7 pairs of strongly marked veins. Scape slender, ridged to the touch, purple-dotted, with base slightly stained purple, up to 15 in (38 cm) tall, well-overtopping leaf-mound, distinctly keel-angled; single foliaceous, navicular-shaped bract, just below inflorescence, green. Flower bracts navicular, pointed, shiny, near-white, purple-tinted. raceme distinctly capitate. Pedicels pale purple. Flowers bell-shaped, lilac, suddenly contracting into narrow tube. Anthers white. Blooms mid-summer and remontant on mature plants in hot climates.

Habitat: Korea: Zennan, Mt Hakuunsan; Japan: Mt Tiisan, Honshu, Mt Rokko, Kyushu, especially on limestone.

Illustr.: Jour. Fac. Sci. Tokyo, Botany 5, pl. 103–104, 1940.

Sub-section spicatae

H. minor   (Baker) Nakai (1911)
Funkia ovata var.   Baker (1871)
   minor   Japanese name:
                             Keirin-giboshi

H. minor is often erroneously thought to be a diminutive form of *H. ventricosa* (syn. *H. ovata* [Sprengel]) and is then referred to as '*H. ventricosa* 'Nana'. Fujita cites *H. longipes* as a synonym of *Funkia* var. *ovata*, but this must be an entirely different plant. The epithet 'minor' is misapplied to several hosta species and cultivars and seems to be used for any small hosta where the true identity is in doubt. For example the name *H. minor* 'Alba' (p. 46) is still in common use for what is correctly *H. sieboldii* 'Alba'. Presumably this has been confused with *H. minor* forma *alba* (Maekawa), the rare white-flowered form of the species *H. minor*.

*

Small, with creeping rhizomes. Late to sprout

in spring. Petiole broadly channelled, winged, up to 3¼ in (8 cm) long. Broadly ovate or cordate, base often truncate, decurrent; margin undulate, apex acuminate; 3½ in (8.5 cm) long × 2¼ in (5.5 cm) wide; glossy mid-to-dark green, with 5–6 pairs of veins. Ridged scapes well-overtopping leaf-mound, up to 24 in (60 cm), longitudinally lined. Bracts obliquely spreading, rather firm, flat, green and purple, gradually withering after anthesis. Flowers funnel-shaped, mauve, up to 10 per raceme. Anthers yellow. Blooms mid-summer. Sometimes remontant.

Habitat: Mainly mountains of Korea.

Illustr.: Gentes Herbarum 2: 140, 1930. *Jour. Fac. Sci. Tokyo*, Botany, 5, 419, 1940.

Forma *alba* (Maekawa) has white flowers.

H. venusta   Maekawa (1935)
                             Japanese name:
                               Otome-giboshi

Young plants of *H. minor* (Nakai) are often confused with *H. venusta* and this not surprising since they belong to the same section and have many points of similarity. *H. venusta* is one of the best-known and most attractive of the dwarf hostas and is used extensively as a rock-garden or peat-bed plant. It would look equally good in a sink-garden planting. The flower scapes, topped with charming bright mauve flowers, stand high above the leaf-mound. The plant now in circulation as *H. venusta* 'Variegated' lacks the ridged scapes that are characteristic of the species and so is presumed to be a hybrid. *H. venusta* is the pod parent of many cultivars, some of which are described in Chapter 4. Its name alludes to Venus and means beautiful, graceful, charming. It is very variable in the wild.

*

Small, clump-forming. Rhizomes slender, creeping. Petiole slender, narrowly channelled, winged, dotted pinkish-purple on lower portion, up to $\frac{3}{4}$ in (2 cm) long. Blade cordate, ovate, base rounded and contracted down, margin slightly undulate or flat, slightly serrated to the touch; acuminate, decurrent; up to $1\frac{1}{4}$ in (3 cm) long × $\frac{3}{4}$ in (1.7 cm) wide; upper surface dull mid-to-dark olive-green, stiff, membranaceous, shiny underneath with 3–4 pairs of thin veins. Scape solid with thin grooves, slender, poised almost vertically, up to 14 in (35 cm) tall. Flower bracts small, green, navicular, not-opening. Flower cluster spaced along upper half of scape and not bunched as in Section Capitatae. Flowers funnel-shaped, widely expanded, light-violet, 4–8 per raceme. Anthers short, oblong, dark-purple. Sometimes remontant. Fragrant forms are said to exist. Flowers from late mid-summer until the first frosts.

Habitat: Korea, Cheju Island; Japan: southern provinces, probably imported.

Illustr.: Jour. Fac. Sci. Tokyo, Botany, 5, 415, 1940; *Hosta: The Flowering Foliage Plant,* Fig. 5.

c) SECTION ARACHNANTHAE

H. jonesii	Chung (1989) *Mo. Bot. Annals*
H. laevigata (Schmidt (invalid)	Korean name: *Tadohae-bibich'u*

Named in honour of Dr Samuel B. Jones, professor of Botany, University of Georgia, USA by Myong Chung, a graduate student researching population genetics and systematics of the Korean hostas for his Ph.D dissertation. Both this and *H. yingerii* have been grown in Professor Jones' garden for several years and he claims that they clearly represent new species.

Found in Kyeongsang Nam Do, Namhae Gun, Namhae Island, Mt Kumsan, growing in the shade of pine and oak forest on rocky, but humus-rich soil, near the coast.

With other new species, *H. jonesii* was collected during an expedition led by Barry Yinger under the auspices of the Friends of the US National Arboretum for the National Arboretum in late summer and autumn 1985.

It was originally placed by Chung in Maekawa's section Tardanthae but was anomalous there. It seems preferable to make it the type plant of a new section, Arachnanthae, as proposed by Schmidt.

*

A small, very variable species. Short, creeping rhizome. Leaves erect, spirally arranged at the base of the stem. Petioles $1\frac{1}{2}$–5 in (4.5–13 cm) long, dotted purple, slightly winged; blades lanceolate or elliptic, $2\frac{1}{2}$–5 in (6–13 cm) long × 1–2 in (3–5 cm) wide, dull dark green, slightly rigid, obtuse or acuminate at apex; 5–7 pairs of veins, glabrous on underside, margin thinly white-lined.

Scapes ascending obliquely, $13\frac{1}{2}$–$23\frac{1}{2}$ in (35–60 cm) tall, dark greenish-purple, purple-dotted at base, 2 lanceolate foliaceous bracts below inflorescence. Flower bracts acute, navicular, green, not open at anthesis, relatively persistent after flowering, pedicels obliquely ascending, whitish-green, minutely purple-dotted.

Flowers quite distinct, spidery (hence proposed new sectional name), lobes, pale mauve, stamens protruding conspicuously. Upper portion of tube campanulate. From 3–20 flowers on each raceme. Flowers late summer, mid-August to early September. Capsules cylindric.

Habitat: Rocks near coast or at the coast, (some in full sun). On rock faces, (often wet) in the mountains.

Illustr.: None available.

H. yingerii	Jones (1989) *Mo. Bot. Annals*

Found in 1985 on islands off south-west coast of Korea by Barry R. Yinger, formerly curator of the Asian Collection at the US National Arboretum, Washington, DC and given to Professor Samuel Jones of the University of Georgia, to be included in the research work of

Haynes Curry who is currently working on hostas for his doctorate.

*

Smallish hosta with leaf mound almost flat on the ground. Short petioles. Differs from *H. jonesii* in having ovate, lustrous leaves and delicate racemes of flowers spread around the central axis of inflorescence; decurrent, flat bracts, relatively longer pedicels, and very protruding stamens.
Illustr.: Hosta: The Flowering Foliage Plant (Pl. 11).

Subgenus *Giboshi*   Maekawa 1938

(a) SECTION HELIPTEROIDES

H. montana	Maekawa (1940)
H. sieboldiana Engler	Makino (1910)
H. fortunei var. gigantea	Bailey (1930)
H. cucullata	Koidzumi (1936)
H. fortunei	Maekawa (1938)
	Japanese name: *Oba-giboshi*

A curious situation exists with regard to *H. montana* in that it is the commonest hosta in the wild in Japan, and also a name, as originally published, of uncertain application and by the ICBN rules invalid. The consensus is to use the name anyway at present, following Maekawa (1940) (p. 51).

H. montana has a very wide distribution. Many new forms, some with unusual and attractive variegation, are currently being collected, mainly on the slopes of Mt Fuji by Japanese hosta specialist nurseryman Kenji Watanabe.

One of the reasons why there is so much

Fig. 7 *H.* 'Birchwood Parky's Gold'
Description on p. 117

difficulty in tying this hosta down to a proper description is its very wide natural distribution and its variability, producing forms ranging from shiny-green to mealy, glaucous grey.

Forms of *H. montana* have been grown in British, European and American gardens for many years. It is a majestic, extremely large and dominant garden plant which looks at its best when grown as a single specimen in a courtyard planting (p. 138).

*

Allowing for wide distributional differences, *H. montana* has thick, bristly rhizomes. The leaf is erect, spreading from the base, inner leaves appearing pendant. Petiole channelled but wide open, green with dull purple spots on both sides at the base, up to 17 in (42 cm) tall. Blade widely ovate, ovate-oblong or ovate-cordate, $9\frac{1}{2}$ in (24 cm) long × $5\frac{1}{2}$ in (14 cm) wide, cuspidate, base shortly cordate or straight, margin occasionally shallowly undulate; upper surface ranges from shiny dark green to glaucous, pale green, dull below with 10–13 pairs of closely spaced veins, deeply impressed. Scape erect, well above leaf mound, solid, round, green, often purple-dotted at base, foliaceous bracts above the centre. Flower bracts spreading, lanceolate, acuminate, remaining in position after anthesis, although withered. Funnel-shaped flowers are dirty mauvish white with bluish-purple anthers. Capsules thick and linear-oblong, Flowers mid-summer. *Habitat*: Southern Hokkaido, central and northern parts of Honshu.
Illustr.: Jour. Fac. Sci. Tokyo, Botany, 5, pl. 21–3; *Hosta: The Flowering Foliage Plant*, Fig. 9.

H.m. forma *aureomarginata* (Maekawa 1940). Leaf blades narrower than type until well-established; blade widely and irregularly margined bright golden-yellow; margins shallowly undulate. A most striking and desirable garden plant, making it one of the most sought-after hostas. It is slow to establish and therefore still rare in cultivation. One of the first to sprout in

spring so is often damaged by early frosts or slugs, if pellets have not been applied in time. It can be difficult to site in a border with other perennial plants because it is so dominant.

H. 'Emma Foster' is a golden-leafed sport of poor constitution, raised from tissue culture.

Var. *praeflorens* (Maekawa 1940) differs from the type in that the flowers open a month earlier and in the more rigid leaves and the thinner corollas.

Var. *liliiflora* (Maekawa 1940). Petioles widely grooved, winged at base, 6–9 pairs of veins. Inflorescence 12 in (30 cm) long, lax after flowering. Anthers near-white, dull purple margin.

Var. *transiens* (Maekawa 1940). Removed from synonymy by Maekawa(1969) p. 52.

The following forms have no botanical standing but are circulating in the nursery trade in Japan and the United States under the following Japanese names. Only recently collected in the wild is *H. montana* 'Fuji no Setsuki', ex Mt Fuji with clear white central variegation, striped and speckled, apex of blade acute, very twisted. 'Aurea', ex Mt Fuji, has leaves yellow, becoming green towards mid-summer. A form with flowers in whorls up the scape is recorded and also one with narrow leaves, both from Mt Fuji and both without names.

H. crassifolia   Araki (1943)
Japanese name:
Atsuba-giboshi

This species was named and published by Araki with a partial description in 1943, but only he recognized it as a species. It is more likely that it is merely a botanical variety of *H. montana*. It differs from *H. montana* in the ovate-oblong leaves, decurrent along the petiole for most of its length. Blade of very thick substance, intensely green on upper surface. Perianth tube is thinner. Not seen outside Japan.
Habitat: Mt Ibukiyama, Shiga Prefecture.
Illustr.: None available.

H. elata	Hylander (1954)
H. fortunei var. *gigantea*	Bailey (1930)
H. fortunei 'Robusta'	Bailey (No Japanese name)

Almost certainly a complex of garden hybrids grown in Europe since the mid-1860s. Possibly derived from *H. crispula, H. sieboldiana* and the *H. fortunei* complex to which it bears a great resemblance. It may, of course, be the same plant as Siebold's *Funkia sinensis* (Hensen 1963) but there is no scientific proof. Or it may embrace some clones of the very variable *H. montana*, the most common hosta in the wild in Japan and the one which Siebold is most likely to have brought back to Europe in one form or another.

Several different clones of *H. elata* were grown for some years at the Case Estates of the Arnold Arboretum of Weston, Massachusetts under different numbers, all from material supplied to Mrs Frances Williams by Dr Hylander from plants grown for many years in Swedish gardens. For discussion of the *H. montana/elata* question see p. 51.

*

H. elata is a large hosta forming dense clumps. Many different clones exist and there are marked differences in the shapes and sizes of the leaves. Some have very undulate margins resembling *H. crispula*, yet others are nearer the old *H. fortunei* × *H. sieboldiana* hybrid which has been growing in European gardens for many years. Petiole slightly glaucous, deeply channelled, winged in upper half, up to 12 in (30 cm) long. Blade ovate, cordate towards base, margin conspicuously undulate in some clones, apex acuminate, 12 in (30 cm) long × up to 7 in (18 cm) wide, matt, mid-to-dark green, slightly glaucous above and below, with 9–11 pairs of veins. Scape up to 60 in (150 cm), glaucous green, occasionally bearing an isolated foliaceous bract below the centre. Flower bracts horizontally spreading, pale greenish-yellow.

Lax raceme of funnel-shaped pale-mauve to milky-white flowers; anthers pale yellow with distinct grey margin. Capsules cylindrical. Flowers mid-summer.

Habitat: European gardens and nurseries since 19th century.

Illustr.: Acta Horti Bergiani, 16, pl. 6, 10–12, 1954;

Journal of the Royal Horticultural Society 81:f.130, 1956.

H. fluctuans	Maekawa (1940)
	Japanese name:
	Kuronami-giboshi

This species is rarely grown outside Japan but seed-raised plants are now being grown in Britain and West Germany. These may or may not be true to type. My description follows Maekawa (1940).

*

Plant of medium size. Rhizomes thick, compact, clustered. Petioles thick, deeply channelled, 9½–20 in (34–50 cm) long, glaucous, green with dull-purple base. Blade exaggeratedly fluctuating, erect or spreading or horizontal to long, tapering to the apex which can twist through 180° (this characteristic not observed on plants of *H. fluctuans* growing in the US), 8–10 in (20–25 cm) long × 4 in (10 cm) wide, dull, dark yellowish-green above, smooth and glaucous-grey below, with 9–10 pairs of veins. Scape strong, leaning, grey and mealy, reaching well above leaf-mound to 40 in (102 cm). Flower bracts long, scarious, yellowish-green, remaining on scape after anthesis. Scape leans more pronouncedly with the weight of the fruit. Flowers funnel-shaped, up to 2½ in (6.5 cm) long, light violet; anthers pale yellow. Flowers after mid-summer. Well fruiting. Capsules long, greenish-white.

Habitat: Honshu mountain areas.

Illustr.: Jour. Fac. Sci. Tokyo, Botany, 5, 358, 1940.

H.f. forma *parvifolia* (Maekawa 1940) has

smaller leaves with less undulation. Rarely cultivated.

H.f. 'Variegated' has leaves much larger than the type, broadly and regularly margined creamy-white to yellow, but may not have been correctly identified. Nevertheless one of the most striking and desirable of the newly introduced variegated-leafed hostas. It is known as *Sagae giboshi* in Japan. (Names not validly published.)

H. crispula	Maekawa (1940)
Funkia marginata	Sieb. nomen nudum
Hosta latifolia var. *albimarginata*	Wehrhahn (1936)
Hosta fortunei marginato-albo	Bailey (1930)
Hosta 'Thomas Hogg'	misapplied
	Japanese name: *Sazanami-giboshi*

There is more confusion over the identity of this hosta than of almost any other – at least in the minds of most gardeners. This is because if it is not grown in perfect conditions, it does not produce its characteristic pronouncedly undulate leaves and is, therefore, often mistaken for either *H.* 'Thomas Hogg' or *H. fortunei* 'Albo-marginata' which is erroneously known in Britain as *H.f.* 'Marginato-alba'. The differences can be seen if petiole cross-sections are compared. In *H. crispula* the petioles are much narrower and more deeply channelled. When well grown there is no mistaking it. It is a magnificent garden plant. It forms a low, wide mound, 26 in (102 cm) across, and the long tapering leaves are undulate across their whole width, not merely along the margins. It is one of the earliest of the white-margined hostas to flower. Being an early flowerer is probably the reason that no hybrids seem to have been raised from it as there are virtually no other hostas in flower at the same time.

H. crispula is one of the few hostas which needs cossetting to perform well. It is susceptible to damage by strong winds and sunlight and it is easily frost damaged.

H. crispula was one of the first hostas to reach Europe. It was imported into Holland by Siebold in 1829. It was described and published by Maekawa (1940) from material cultivated in Japanese gardens and there is only one reference to an exact location in the wild. Plants of *H. crispula* were brought to Europe on many occasions by Siebold's collectors. J. Pierot, one of Siebold's collectors, notes that one plant was collected 'In valle montis Fawara Saka principatus Fizen, ins. Kiu Siu' and placed in the Herbarium at Leiden.

It is one of the several hosta species first known and described in its variegated form, the green-leafed form being discovered later. *H. crispula* is obviously closely related to *H. elata*, a probable hybrid of European origin, and since *H. crispula's* introduction predates the existence of *H. elata* (p. 51), it is possible that it may be one of the parents or ancestors of that hosta, even though the flowering times do not appear quite to coincide, with an *H. fortunei* or an *H. sieboldiana* as the other. The green-leafed form, *H.c.* forma *viridis* (Brickell 1968) is little cultivated. The yellow-blotched *H.c.* f. *lutescens* was first offered by Siebold's nursery in 1876. The basionym is *Funkia marginata* var. *lutescens* listed in *Sieb. Prix-cour.*, 1876, 1, (Hensen 1963). It is also listed in *Sempervirens* 20; 532, 1893 as *F. (albo)-marginata* var. *lutescens*. Hylander named it *H. crispula foliis maculatis* hort. as he considered it to be a cultivar. It has been variously described in the literature as being 'spotted' or 'viridescent'. This form of unstable variegation normally fades with the onset of the warm weather.

*

Medium to eventually large hosta. Slow-growing but ultimately making very big, dense clumps. Leaves arched, curved, spreading and hanging down towards the tip which almost touches the ground on the outer leaves. Rhizomes thick, much branched. Petiole shortish – 11 in (28 cm), deeply U-shaped on young plants, petioles on outer leaves of mature plants less deeply U-shaped. Blades strongly undulate, ovate-lanceolate, long, tapering, often twisted through a full 180° towards the apex, and pendant towards the tip, base blunt or cordate, often rolled round the petiole, widely and irregularly margined in clear white; margins continuing narrowly down outer edges of petiole; 7 in (18 cm) long × 4 in (10 cm) wide in Britain and Europe (probably larger in America); matt, olive-green above, glossy below, with 7–9 pairs of deeply impressed veins. Scape up to 36 in (90 cm), slender, round, green, overtopping leaf mound. One or more foliaceous bracts. Flower bracts small, spreading, whitish to pale-green. Flowers 30–40 per scape, sparse lower down, funnel-shaped, pale lavender (same colour in bud); anthers large, light purple, becoming yellow at dehiscence. Seed is freely formed. Capsules cylindrical. Flowers mid-summer.

Habitat: Cultivated in Japanese gardens.
Illustr.: Acta Horti Bergiani 16, pl. 13, 1954
Hosta: The Flowering Foliage Plant, Figs. 8 & 21.

H. nigrescens	(Makino 1902 Maekawa 1937)
H. sieboldiana var. *nigrescens*	Makino (1902) Japanese name: *Kuro-giboshi* (meaning black or dark hosta)

A species until recently not much cultivated outside Japan where it has been grown in gardens for a long time. Although a northern Japanese species it has been much prized for planting round shrines, along with *H. sieboldiana* and other glaucous, large-leafed hostas, and because of this it is now cultivated all over Japan (I am indebted to George Schmid for this information).

H. nigrescens is remarkable for the blackness of the emerging shoots and it is perhaps for

these that Makino named the plant. It is distinct, even in its section, on account of the intense ash-grey pruinosity covering the entire plant when the leaves first unfurl, disappearing later; the poise of the leaves, their smoothness and also for the height of the scapes (up to 7 ft (2.1 m)) zig-zag in outline in young plants but ramrod-straight in mature plants. A possible parent of the *H. fortunei* complex (the tall scapes of *H.f.* var. *hyacinthina* are an indication of this); a possible parent of *H.* 'Krossa Regal' and perhaps related to *H.* 'Tall Boy' which also has exceptionally tall scapes.

A plant of beautiful proportion and exquisite colouring that deserves to be better known.

*

A large hosta forming an open mound 20 in (52 cm) high × 24 in (60 cm) wide. Sprouting shoots pruinose charcoal black. Rhizomes thick with sharp bristles. Petioles thick and straight, up to 20 in (52 cm) tall, glaucous to pale yellow-green, deeply channelled, winged in upper half and spotted purple. Leaves erect and spreading. Blade wide, ovate to elliptic, slightly dished, 11 in (28 cm) long × 7 in (18 cm) wide, cordate at base and rolled so as to form a funnel downwards to the decurrent petiole; upper surface leathery; with 8–13 pairs of feintly impressed veins, dark green, dusted ashen grey-green at first, turning glossy dark green later. Scape straight or arching and undulate in young plants in some clones, glaucous grey-green, held high above the leaves, up to 5 ft (150 cm), and exceptionally 7 ft (210 cm) tall in Britain and more often in American gardens, bearing 2–4 spreading foliaceous bracts. Flower bracts greenish-white with purple margins, glaucous outside. Flowers 30–40 per raceme, buds pale-purple, turning milky-white on opening. Flowers densely clustered, flaring, funnel-shaped, 2 in (5 cm) long. Seed capsules broad, oblong, sometimes sterile. Flowers late summer.

Habitat: Mountains of northern Honshu, but mainly cultivated in Japanese gardens and shrines.

Illustr.: Jour. Fac. Sci. Tokyo, Botany, 5, 323, 1940.

H.n. var. *elatior* is much larger in all respects with no pruinosity.

H. sieboldiana	(Hooker) Engler in Engler & Prantl, *Natürl. Pflanzenfam.*, 11:5; p.40, 1888
Hemerocallis Sieboldtiana	Loddiges in *Bot. Cab.* 19, no. 1869, 1832
Funkia sieboldiana	Hooker in *Curtis' Bot. Mag.* LXV: t. 3663, 1838 (as Funkia): *Kunth, Enum.* Pl.IV: 592, 1843
Funkia sieboldi	Lindley in Edwards' *Bot. Reg.* 25, tab. 50, 1839
Saussurea Sieboldiana	(Hook.) O. Kuntze, *Rev. Gen.* II, 1890, p. 714
Niobe sieboldiana	(Hook.) Nash in *Torreya* XI:6, 1911, *Libertia*
Hosta glauca	Stearn in *Gard. Chron.*, ser. 3, XC:88, 1931, (in part)
	Japanese name: *To-giboshi* (Maekawa ex Iinuma & Iwasaki)

H. sieboldiana is regarded in Britain, Europe and America as the archetypal hosta. It and its forms are the best known, certainly the most popular and most probably the most garden-worthy plants in the genus; with their stands of heavily puckered, glaucous, heart-shaped leaves they are what most people expect hostas to be. Plants of this species are, however, often too large for today's smaller gardens but the Tardi-ana group, of which *H. sieboldiana* is one parent, contains a number of similar but smaller substitutes.

H. sieboldiana is a variable plant in the wild, its leaves ranging from shades of green to glaucous-blue. It has a very limited wild distribution in Japan, being confined to Yamagata and Aomori. It is prodigal with seed and many plants sold as the species are in fact seed-raised and differ greatly. Tissue-culture, however, makes large quantities of true plants available.

The history of *H. sieboldiana's* importation first into continental Europe in 1829 and thence to Britain, and the subsequent muddle because of the similarity of the name with *H. sieboldii*, is documented on p. 41.

H. sieboldiana is easy to grow but does best when given a sheltered, semi-shaded position with ample moisture at its roots. It is not one of the first to sprout in spring, although not nearly as late to appear as its relative, *H. tokudama*. If it is given a regular diet of animal manure and non-organic fertilizer, it will produce leaves up to 18 in (46 cm) long. It will also withstand a good deal of drought but luxuriant growth cannot be expected in drought conditions. It is not a good hosta to grow under trees as its slightly cupped and ridged leaf-surfaces collect and hold any falling debris. Too much water falling directly onto the leaf surface will mark the waxy glaucous bloom giving a patchy appearance. It looks best when grown as a specimen plant in a paved courtyard or *en masse* in a waterside planting (Fig. 35), as a foil for astilbes, ferns and *Primula florindae*. It is much used as a plant for flower arrangements as both the leaves and flowerheads are very distinctive in outline. The seedheads remain intact on the scape long after the leaves have gone, wan, sere and slightly ghostly, providing interest in the winter.

*

A robust, strong-growing, clump-forming perennial of almost shrub-like proportions. Rhi-

Fig. 8 *H. crispula* forma *viridis*
Description on p. 72

zomes short, thick and erect and tending to lift above the ground after a few year's growth. Sprouting shoots greyish-mauve. Petiole up to 24 in (60 cm) (longer than blade on mature plant), deeply grooved with often only $\frac{1}{4}$ in (1 cm) between the two edges, erect, light-green, almost white at base. Blade almost at right-angles to petiole, nearly parallel with the ground, ovate-cordate to nearly orbicular, base distinctly cordate, margin almost flat, apex acute or coming to an acuminate tip; up to 18 in (50 cm) long × 12 in (30 cm) wide; rigid, thick, rugose, matt, glaucous blue-grey to green above, paler blue-grey to green below and covered on both surfaces with a pruinose bloom. Veins 14–18, deeply impressed, the outer veins strongly curved. Scape obliquely ascending, 24 in (60 cm) tall, glaucous, whitish-green, sometimes sparsely dotted purple. Raceme 10 in (25 cm) long. Foliaceous bracts distinctly elongated. Flower bracts single, scarious, concave, becoming flat at apex at anthesis. Flowers densely packed on raceme; funnel-shaped, not widely opening, $1\frac{3}{4}$ in (5.5 cm) long, very pale, mauvish-grey, fading to almost white; occasionally white; anthers yellow. Large capsules, well fruiting. Flowers early to mid-summer, usually June. Very occasionally remontant.

Habitat: Yamagata and Aomori.

Illustr.: Acta Horti Bergiani, 16, pl. 3 & 4, 1954.

Jour. Fac. Sci. Tokyo, Botany, 5, 370, 1940.

H.s. var *hypophylla* (Maekawa 1940) (syn. *H. glauca* [Siebold ex Miquel]) has shorter scapes and almost round leaves. It is sometimes described as a non-flowering form which is incorrect but the flowers do not show above the leaf mound.

Var. *amplissima* (Maekawa 1940) produces dense clumps of very large leaves; blades elliptic-ovate, shortly cordate at base, suddenly acute at apex with slightly undulate margins; scape arching so that raceme is held almost horizontal with the flowers pendant.

Var. *mira* (Maekawa 1938) is a Japanese garden form which in America is the largest of all the cultivated forms – a plant of almost tropical vigour and appearance. In Britain it is less robust but still very fine. Differs from other forms in its more pointed leaf apices, longer, thicker, bending and more glaucous petiole and in the purple flush on the outside of the tube which is otherwise white. Scape stands slightly higher above the leaf mound than in other forms. *H.* 'Semperaurea' (p. 54) is now considered to be the golden form of this hosta.

Var. *elegans* (Hylander 1954, an Arends [1905] hybrid) is a superlative foliage plant in a stout, architectural idiom. It differs in its greater glaucousness, its more puckered leaf blade (resembling gently starched seersucker), and in the almost opalescent quality of its pale lilac flowers. The garden merit of this plant is indisputable: its taxonomic status is less clear. It has never been found in the wild but was discovered in Georg Arends's nursery in Wuppertal, Germany, in about 1905 (p. 40). It does not come true from seed and, if Arends is correct, it is a hybrid between *H. sieboldiana* and *H. tokudama.* Selfed seedlings run the whole gamut of intermediates from one species to the other. Var. 'Aureo marginata': see under *H.* 'Frances Williams' (p. 53).

Illustr.: Hosta: The Flowering Foliage Plant, Fig. 22.

Var. 'Aurea'. Origin unknown, but presumably arose as a sport. See also under recently registered name *H.* 'Golden Sunburst' p. 112.

Var. *cucullata* Sieb. See p. 41.

There are many other forms and seedlings emanating from *H. sieboldiana,* too many to list, with yet more named and registered each year. In Chapter 4 there is a selection.

H. tokudama	Maekawa (1940)
Funkia fortunei	Anonymus (Sieb. Cat. 1870/71)
Funkia sieboldiana fortunei	Regel (Gartenflora 1876)

Funkia sieboldiana (Hook.) var *condensata*	Miquel (1867)
H. sieboldiana f. *fortunei*	Voss (Vilm. Bluemeng. 1896)
H. sieboldiana var. *glauca*	Makino (1902) & Matsumura (1905) Japanese name: *Tokudama-giboshi*

H. tokudama is, in spite of its singular slowness to increase, one of the most beautiful and most sought-after hostas. It is, in effect, similar to a dwarf version of the best form of *H. sieboldiana* (e.g. var. *elegans*) though the leaf is much more cupped, and it is ideally suited to small gardens.

It was introduced to Europe by Robert Fortune in 1862, probably from material of Japanese garden origin. Makino (1902) and Matsumura (1905) considered this species to be *H. sieboldiana* var. *glauca*, and its position in the genus *Hosta*, is still in dispute today. (Hensen 1985). Originally considered a species, it is now thought that it may be a garden hybrid rather than a true species.

In common with *H. sieboldiana* it is best grown in open ground away from overhanging trees (p. 74). It is the most vividly glaucous-blue of all hosta species and the leaves are of very heavy substance and heavily textured, as well as deeply cupped.

*

A small to medium-sized hosta of compact habit and slow growth. Blade held stiffly and sharply away from the petiole. Petiole broad, grooved, glaucous-blue, up to 8 in (20 cm). Blade cordate to orbicular, abruptly tapering to a firm and distinct apex. Margin flat. Upper surface rugose, intensely glaucous-blue, well marked with 13 pairs of veins; veins prominent on underside which is less glaucous. Scape equal to, or slightly taller than, leaf-mound, although occasionally shorter, approximately 12 in (30 cm) tall, stiff, thick, glaucous. Flower bracts oblong/lanceol-ate, rigid, tinted or suffused purple. Flowers round, bell-shaped, not opening wide, very pale dirty mauvish-grey, almost white, turning down or facing sideways on raceme. Anthers yellowish. Sets little seed, although many hybrids have been raised from it. Capsules short, clustered. Flowers late mid-summer.

Habitat: Maekawa retracted his original statement that living material had been seen on Mt Akiwa, Tottori Prefecture. All the plants of *H. tokudama* he actually saw were in Hortus Kikuchi, Dr Kikuchi's own garden, and were identified only by numbers.

Illustr.: Jour. Fac. Sci. Tokyo, Botany, 5, 37; *Hosta: The Flowering Foliage Plant*, Fig. 20.

The following forms are often listed as cultivars although correctly botanically described.

H.t. forma *flavo-circinalis* (Maekawa 1940). Larger than all other forms of *H. tokudama*, blade more ovate and less acutely tapered at apex. Margin clear bright primrose-yellow but tending to turn brown early in autumn. Even slower to establish than the type. In Britain it has arisen as a sport from *H.t.* f. *aureo-nebulosa*.

H.t. forma *aureo-nebulosa* (Maekawa 1940). Blade similar to type, cloudy-yellow centre, irregularly margined glaucous-blue. (Many so-called improved forms exist, e.g. 'Blue Shadows', but this form is variable and unstable.)

Illustr.: Hosta: The Flowering Foliage Plant, Fig. 10.

H.t. forma *flavo-planata* (Maekawa 1940). Blade similar to type, centre deep-yellow, narrow, glaucous-blue margin. A plant of garden origin – perhaps a sport of *H.t.* f. *aureo-nebulosa*.

Gold-leafed sports arising from *H. tokudama* are grouped under *H.* 'Golden Medallion' (q.v., p. 172), except for those arising from *H.t.* f. *flavo-circinalis* which are now known as *H.* 'Gold Bullion' as they also do not follow the leaf size and shape of the type. The best of the named hybrids are listed under their cultivar names in the next chapter.

H. fortunei   (Baker) Bailey 1930
Funkia fortunei   Baker (1876)
Niobe Fortunei   Nash (1911)
                         Japanese name:
                              Renge-giboshi

Named for Robert Fortune, the intrepid plant collector, the scientific identity of this hosta remains a mystery. The problem is that Baker, who first validly published the name in 1876, based his description on living material which can no longer be traced and he left in the Kew Herbarium a specimen now in such a poor state of preservation that it is no longer possible to deduce its actual identity (p. 44). The hostas widely grown in gardens under this name are an assemblage of plants that are substantially similar to each other, probably hybrids between several species in the section Helipteroides (which they closely resemble) and hostas from another section as yet undetermined. They are grouped for convenience under the umbrella name *H. fortunei* and were divided into botanical variants and clones by Hylander (1954) from plants which it is now believed were entirely of European origin. Some of these have recently been registered by The American Hosta Society and given cultivar status. Sports from these botanical variants have arisen in cultivation over the years and these also have cultivar names. A selection of the most garden-worthy of these is included in the cultivar chapter.

Hostas of the *H. fortunei* group have never been described by Japanese botanists and it is believed that they do not even exist in Japan other than as plants imported from European gardens and nurseries, although the plant described as *H. fortunei* (Baker 1876) is only possibly the one brought back from Japan by Robert Fortune; in fact there is confusion here since Fortune's hosta is now properly *H. tokudama* (p. 44).

*

Large, robust plants, making extensive clumps of striking foliage. Petioles glaucous or very glaucous, up to 14 in (35 cm) long, deeply channelled and distinctly winged. Blade cordate or cordate-ovate, sometimes almost lanceolate-ovate; apices gradually or sharply pointed, margins somewhat undulate, sometimes flat, decurrent; in some cases surfaces slightly rugose; up to 12 in (30 cm) long × 7½ in (20 cm) wide; dull mid-to-dark green, sometimes glaucous above and usually glaucous below, with 8–10 pairs of veins. Scape well overtopping leaf-mound, up to 30 in (90 cm), usually glaucous and in most cases bearing at least one well-developed foliaceous bract. Flower bracts glaucous, green, sometimes suffused violet, navicular, producing a cone-like inflorescence before anthesis. Flowers funnel-shaped pale mauve to light violet, with narrow lobes. Anthers bluish-purple with irregular pollen. Fruit usually abortive. Flowers late summer.

Habitat: Gardens in Britain, Europe, the United States and Japan.

Illustr.: None known.

H.f. var. *rugosa* (Hyl). Similar to var. *hyacinthina* (Hyl.) and var. *stenantha* (Hyl.), but blades smaller and more glaucous than in var. *stenantha*. Blades markedly rugose between the veins; matt and pruinose on upper surface, less so than in var. *hyacinthina* with 8–9 pairs of veins. Scape shorter than var. *hyacinthina* but still well above leaf mound.

Illustr.: Acta Horti Bergiani, 16, pl. 5, 1954.

Var. *stenantha* (Hyl.), little grown or offered commercially now. Blade mid-to-dark green, shiny on upper surface if grown in sun, glaucous in shade; slightly pruinose below; shallowly undulate margin, 9–10 pairs of veins. Shortish scape with only the inflorescence appearing above the leaf-mound. Has been confused with *H. ventricosa*.

Illustr.: Acta Horti Bergiani, 16, pl. 7, 1954.

Var. *hyacinthina* (Hyl.) (*H. glauca* misapplied). Still very much in evidence in gardens, though seldom offered commercially now. Blade glaucous, greyish-green above, bluish-grey below;

slightly leathery texture with very shallowly undulate margin; easily distinguished from var. *rugosa* by thin hyaline margin, 9–10 pairs of veins. Scape overtops leaf-mound conspicuously in this variant.
Illustr.: Acta Horti Bergiani, 16, pl. 5, 1954; *Hosta: The Flowering Foliage Plant*, Fig. 4. Many unstable variegated forms of var. *hyacinthina* exist, and the variegation of most of them tends to fade at the onset of the hot weather. The best of these are listed in Chapter 4.

Var. *albopicta* (Miquel) Hyl. (*H. fortunei* 'Aureomaculata'; *H. lancifolia* (Thunberg) Engler var. *aureomaculata* (Wehrhahn). Possibly the *Funkia viridi marginata* of Siebold since this plant was sold as 'Viridis marginata' as late as the late 1960s. Leaves unfurl creamy-yellow with irregular mid-to-dark green margins with very fine dark cross-veining. Commonly known in Britain as 'The Chelsea Hosta' because year after year it has been one of the showiest plants at the Royal Horticultural Society's Chelsea Flower Show, and features on many exhibits when its foliage is at its peak in May. In hotter climates, including most of the United States, this variegation fades much more quickly than it does in British gardens and it is, therefore, a much less desirable garden plant in such climates. The blade gradually turns green but a vestige of the variegation is always visible. Blade slightly larger and thinner in texture than those of the foregoing variants, with 8–9 pairs of veins.
Illustr.: Acta Horti Bergiani, 16, pl. 6 & 8, 1954. Thomas, G.S., *Plants For Ground Cover*, Fig. XIA, 1970.

H.f. var. *albopicta* forma *aurea* (Wehrhahn 1936). Forms with yellow leaves of varying shapes and sizes exist, but in general the blade is slightly smaller and more elliptic than in var. *albopicta*. Blade ovate-elliptic, slightly acuminate, slightly rugose above, glaucous below, of thin substance, becoming shiny if grown in full sun; lacking green margins; gradually turning pale, dull green during the summer. Will scorch in direct strong sunlight in Britain, especially if droplets of water fall on the leaf surface.
Illustr.: Thomas, G.S., Plants for Ground Cover, Fig. XIB, 1970.

Forma *viridis* (Hyl.). Green-leafed sport of *H.f.* var. *albopicta*.

Var. *obscura* (Hyl.). Probably the nearest to Baker's *H. fortunei* and possibly synonymous with *H. bella* (Wehrhahn) according to Hylander. (*H. bella* is now presumed to have disappeared from cultivation (p. 52). Named by Hylander from material growing only in one Swedish garden. Blade broadly cordate-ovate, 9 in (23 cm) long × 6¼ in (16 cm) wide; 11–13 pairs of veins, shortly acuminate, slightly rugose, dull, dark green on upper surface, slightly glaucous below. Named *obscura* by Hylander on account of its dark green colouring. Scape much taller than leaf-mound. Hardly ever grown nowadays. Much more popular in its yellow-margined form var. *aureo-marginata* (Wehrhahn) (syn. *H.f.* var. *obscuro-marginata* and 'Yellow Edge'), not to be confused with *H.* 'Viette's Yellow Edge' which is a variegated form of *H.f.* var. *hyacinthina*. The adoption of Wehrhahn's name was necessary to allow the listing in the Supplement to the Royal Horticultural Society's *Dictionary of Gardening*, Oxford, 1969, to follow the rules of priority. The American Hosta Society has since registered it with the cultivar name 'Aureo-marginata'. Hylander believed that this plant arose in a Swedish garden, but it is more likely that it arose in Holland or Germany and was taken to Sweden sometime before 1936, when it was described in *Gartenflora*. It was also thought to have arisen in Siebold's nursery a century earlier.
Illustr.: Thomas, G.S., Plants for Ground Cover, Fig. 31, 1970; *Hosta: The Flowering Foliage Plant*, Fig. 1.

H.f. 'Albo-marginata'. This cultivar name for a fine, showy white-edged *H. fortunei* possibly a white-margined form of *H.f.* var. *obscura*, has been registered by the American Hosta Society, partly in order to get round problems presented by the *albo-marginata* and *marginato-alba* epithets, which get used interchangeably because of their similarity.

It could be argued that *H.f.* var. *marginato alba* (Bailey 1930) should be included here, but both Maekawa (1940) and Hylander (1954) regard Bailey's name as a synonym of *H. crispula*. Bailey does not give sufficient information for anyone to be sure which white-margined hosta he means. It may also be that other white-margined hostas have been sold under that name too. A large, robust plant. If grown in shade in good fertile soil with moisture at its roots, it is quite the largest of the *H. fortunei* group. Blade up to 12 in (30 cm) long × 6⅔ in (17 cm) wide. Oblong-ovate, acute with irregular wide margins, sometimes spreading over half the blade, pale-cream at first, becoming pure white; upper surface, dull, matt, mid-to-dark green, thinly pruinose beneath; 7–8 pairs of widely spaced veins. Scape up to 38 ins (95 cm); one foliaceous bract. Flowers funnel-shaped, light violet to mauve. Blooms mid-summer. Most seeds abortive.
Illustr.: None available.

(b) SECTION RHYNCHOPHORAE

H. kikutii	Maekawa (1937)
	Japanese name:
	Hyuga-giboshi
	(taken from
	location of
	discovery)

Maekawa described this plant (*J. Japanese Botany*, 13, 48, Fig. 9, 1937) naming it in honour of his teacher Dr Akio Kikuchi. The Latinate pre-war classic spelling of Dr Kikuchi's name was Kikuti; it was from this that the species name was derived and under the rules of nomenclature the species name must remain as originally cited. *H. kikutii* occurs in climatically mild areas in the mountains and on forest margins. Its distinguishing feature is the beak-shaped flower bud and also the leaning scapes, common throughout this section. The type plant, *H. kikutii*, is the largest of many botanical variants, several of which have only recently been found in the wild and have not yet been analyzed or classified. Some of these have, for the present, been given cultivar names, e.g. *H.* 'Harvest Delight'. The type plant, and *H.k.* var. *polyneuron*, and *H.k.* var. *caput-avis* (now reduced to varietal level) are the only representatives grown in Britain, though var. *polyneuron* is only known there in seed-raised form.

H. kikutii hybridizes freely in the wild with *H. longipes* (their flowering periods overlap) and they grow in the same localities, so there will also be a problem of assigning the interspecific hybrids arising in the wild. It is now understood that many different variegated forms have arisen in the wild but none has, as yet, been botanically described.

Some forms of *H. kikutii* are delightful little hostas suitable for growing in the rock garden or peat bed, where their graceful characteristics can be minutely observed, but other forms are larger and more suitable for woodland or border. On the whole the larger plants are less attractive, though the cultivar 'Green Fountain' is one of the best of the newer green-leafed hostas raised.

*

Open leaf mound, 8½ in (22 cm) high × 12 in (30 cm) wide. Petiole shorter than blade, deeply channelled, green, sometimes purple-dotted at base. Blade ovate-lanceolate, arching, tapering to apex, margin rippled, base narrowly cordate, matt or sometimes glossy bright mid-green above, shiny below, with 7–10 closely spaced, deeply impressed veins. Scapes leaning, rigid, surpassing leaf mound (type plant leans least of all) with sterile foliaceous bracts above the centre, dark purple to green. Flower bract navicular, suffused purple, withering after anthesis. Flowers in dense raceme on long, strong purple-rose-pink pedicels, near-white, trumpet-shaped, flaring, stamens protruding from mouth; anthers bluish. Very fertile.

Habitat: North east from the islands in Osumi Gunto, including Yakushima and Kyushu and Shikoko.

Var. *polyneuron*. Generally a much smaller plant with leaves broader than the type and more, and more closely spaced, veins. Scape leans at 45°. The Japanese name *Sudare-giboshi* alludes to the closely spaced veins which resemble the slats of a bamboo curtain or blind (Schmid 1987). According to Kenji Watanabe, a gold-leafed form of *H.* var. *polyneuron* exists, known as *Kisudare-giboshi*.

Illustr.: Hosta: The Flowering Foliage Plant (Fig. 19).

Var. *pruinosa* is the most erect of the group and has a white pruinose underside to the leaf and petioles.

Illustr.: Hosta: The Flowering Foliage Plant (Pl. 5).

Var. *yakushimensis*, contrary to its name, comes from Kyushu and not Yaku Island. The darkest green of the botanical variants with leaves lying almost flat on the ground. Scape leans very markedly.

Illustr.: Jour. Fac. Sci. Tokyo, Botany, 5, pl. 42–44

Var. *caput-avis* was given species rank by Maekawa but is merely a variant of the type. Different forms occur. The Japanese name *Unazaki-giboshi* means 'the nodding hosta'. Smaller than the type, the scapes bend from the crown and grow parallel to the ground with the flowers flattened on the raceme. The name caput-avis means 'head of a bird', alluding to the pronounced beak-shaped bud surrounded by the sterile bracts.

Var. *tosana* was also classified as a species by Maekawa but Fujita reduced it to varietal rank in 1976 under var. *caput-avis* which it slightly resembles, although it is a larger plant.

Illustr.: Jour. Fac. Sci. Tokyo, Botany, 5, pl. 45–56.

Kenji Watanabe's book on hostas lists a gold-streaked form of *H.k. caput-avis*, forma *variegata*, as Fuiri *Unazaki-giboshi*.

H. shikokiana   Fujita (1942)
Japanese name:
Shikoko-giboshi

This hosta is only recognized as a species by Fujita. It is not known in Britain or Europe. This description is based on that of Fujita. I include it in Rhynchophorae, though it has been placed in Eubryocles by some authors.

*

Small. Rhizome short and erect. Leaves semi-erect, tufted at rhizome. Petiole up to 4 in (10 cm), spotted dark purple at base. Blade ovate, apex gradually acuminate; upper surface slightly lustrous, with 5–10 pairs of veins spaced fairly close together, elevated below. Scape obliquely ascending, well above leaf-mound, up to 16 in (40 cm), dotted purple. Foliaceous bracts just below raceme, ovate-lanceolate. Flower bracts navicular, remaining closed at anthesis and losing their intense green colour before anthesis. Flowers light to mid-purple, stamens slightly protruding. Anthers oblong, purple. Fruit cylindrical. Seed small and light. Flowers late summer.

Habitat: Outcrops along ridge lines of Ishizuchi range and the Akaishi range in Shikoku, Japan.

Illustr.: None available.

H. pycnophylla   Maekawa (1972)
Japanese name:
Noshi Setouchi-giboshi

The specific name means 'leaves clustered together', but just how characteristic this is of the species is difficult to say. Our own plant has proved so slow to increase that this quality is not yet apparent. A graceful and attractive hosta still little known in Britain and Europe and, apparently not much known in Japan either.

Noted for the snow-white, densely pruinose underside of the leaf blade. In spite of the pruinosity it does not seem to be able to tolerate full sunlight. Suitable for the woodland garden, raised peat bed or rock garden it dislikes heavy clay soil. One of the very best of the smaller hosta species. The description given below is based on observations from our own young plant, from information supplied by George Schmid and from Fujita's brief description. *H. pycnophylla* is used by breeders to raise cultivars with ruffled margins.

*

Petiole narrow, deeply grooved, glaucous, suffused dark purple and purple-dotted, up to 4 in (10 cm). Blade ovate-lanceolate, gradually acuminate margins very undulate; up to 8 in (20 cm) long × 4 in (10 cm) wide, upper surface matt, light sage-green, underside thickly coated with white pruina, thin textured, 6–8 pairs of well-marked, widely spaced veins. Scape narrow, dark purple, obliquely ascending, over-tops leaf-mound and bears a small green folia-ceous bract. The developing flower bract is closed (as in *H. kikutii*) and becomes navicular at anthesis. Flowers trumpet-shaped, with colour ranging from light mauve to deep purple, sometimes almost crimson. Stamens protrude from perianth tube. Fragrant forms are said to exist. Blooms late summer and into autumn.
Habitat: Mountain ridges and crags, Oosnima Island, Honshu and a few other islands in the Inland Sea.
Illustr.: None available.

(c) SECTION INTERMEDIAE

H. kiyosumiensis   Maekawa (1935)
                            Japanese name:
                                Kiyosumi-giboshi

Found on Mt Kiyosumi by Professor Nakai. Listed once by a British nursery but not often grown outside Japan. Not a gardenworthy plant on account of the sparsely flowered raceme.

*

A small to medium-sized hosta which achieves a graceful open mound with spreading leaves. Rhizomes thick, fibres of spent petiole persist-ing. Petiole very short, broadly grooved, winged, almost triangular in cross-section, pur-ple-dotted at base, up to $3\frac{1}{2}$ in (8 cm), shorter than blade. Blade ovate-elliptic, $4\frac{1}{2}$ in (11 cm) long × $2\frac{1}{2}$ in (7 cm) wide (similar to those of the *H. fortunei* type); margin undulate, uneven and irregularly, minutely toothed; apex acumin-ate; shiny, yellowish-green above, dull below, of papery texture; with 5–6 pairs of veins. Scapes much taller than leaf mound, up to 15 in (38 cm); purple-dotted at base, some foliaceous bracts. Raceme bears 4–5 flowers which stand out sideways from scape. Flower bracts similar to those of *H. lancifolia*, green, navicular. Flowers white, tubular, bell-shaped; anthers pale, dull purple. Capsules long, cylindrical. Flowers midsummer.
Habitat: Mt Kiyosumi, Chiba Prefecture.

Var. *petrophila* (Maekawa) is a much smaller form with flowers that are often dull blue. Flower tube much narrower. The varietal name alludes to the rock-like habitat that this hosta enjoys.
Habitat: Kinki district, Honshu.
Illustr.: None available.

H. densa   Maekawa (1938)
                    Japanese name:
                        Keyari-giboshi or
                        Odaigahara-giboshi (after the
                        mountain on
                        which it was
                        found)

Not cultivated in Britain and probably not grown at all outside Japan.

*

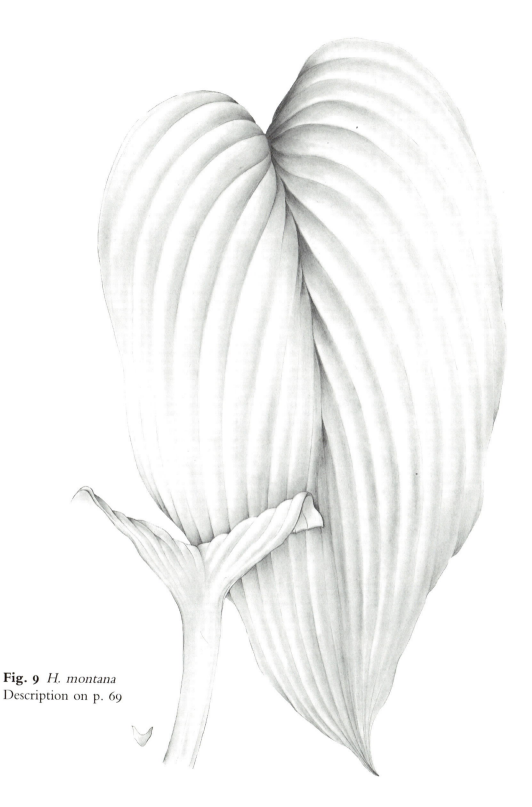

Fig. 9 *H. montana*
Description on p. 69

Rhizome thick, short and bristly. Mound open, leaves spreading. Petiole erect, $8\frac{1}{2}$ in (21 cm) long, widely grooved, spotted dull purple at base. Blade ovate-elliptic, with pronounced undulate margins, base blunt, often convolute, apex acuminate, 8 in (21 cm) long × 4 in (10 cm) wide, similar in shape to that of *H. sieboldiana*; with 8–10 pairs of veins, raised; upper surface dark green, paler below. Scape thick, round, far overtopping leaf-mound, 24 in (60 cm) tall, dull green, spotted dark purple at base. Flowers closely spaced on bent-back raceme; bell-shaped, faded purple. Flower bracts overlapping, navicular or flattened, green, acuminate; anthers long, yellowish white. Flowering time late summer.

Habitat: Mt Odaigahara, Honshu.

Illustr.: Jour. Fac. Sci. Tokyo, Botany, 5, pl. 47, 48, 49.

H. pachyscapa Maekawa (1940)
Japanese name:
Benkei-giboshi

Not grown in Britain and little known outside Japan.

*

Medium-sized hosta with spreading mound, leaves held horizontally. Petiole open-grooved, winged, green, spotted purple at base, up to 10 in (25 cm) tall; blade ovate-elliptic or oblong, acuminate; contracted and obtuse at base and often closed; margin slightly undulate; lustrous deep-green above with 7 pairs of veins; $5\frac{1}{2}$–7 in (14–18 cm) long, $2\frac{1}{2}$–$4\frac{1}{2}$ in wide (6.5–11.5 cm) wide. Scape stout, up to 38 in (95 cm), dotted-purple; sterile foliaceous bracts. Flower bracts imbricated, green with purple margin, lanceolate. Flowers densely clustered, funnel-shaped, pale purple; anthers blue. Flowering time August.

Habitat: Kinki district, Honshu.

Illustr.: None available.

(d) SECTION STOLONIFERAE

H. clausa Nakai (1930), *Bot. Mag. Tokyo*
Japanese name:
Tsubomi-giboshi

Section Stoloniferae is represented only by the species *H. clausa* and its forms and varieties. This section is the equivalent, in the Korean peninsula, of the Section Nipponosta in Japan. It is noted for its very long stolons often as much as 1 ft (30 cm) long. The leaves are held stiffly erect, and the permanently closed flower buds, (one of the main characteristics of the species), have strongly impressed veins. Probably because *H. clausa* is a botanical curiosity rather than an attractive garden plant, it is rarely cultivated in Britain, although it is used as ground cover in parts of the Savill Garden, Windsor Great Park where its stoloniferous habit makes it a useful landscaping plant. *H. clausa* is sterile because it is a triploid (2n = 90). Var. *normalis* is also sterile, but has beautiful, richly coloured flowers and considerable garden potential.

*

Medium-sized, spreading hosta. Rhizome very long, stoloniferous, radiating horizontally from the centre; 8 in (20.5 cm) to 15 in (43 cm), turning white in summer. Inter-nodes up to 1 in (2.5 cm) long, emerging nodes $\frac{1}{3}$ in (2 cm) long, with withering scales and a few fibrous roots. Root-stock erect, thick and very short. Petiole longer than blade in mature plant, up to 6 in (15 cm), deeply channelled, pale green, purple-dotted at base. Blade decurrent, held erect and stiff; arranged in rows 6–8 in (15–20 cm) high, lanceolate, apex acuminate, margin flat, mid-olive-green, slightly paler and shiny below, with 4–5 pairs of veins. Scape 15 in (38 cm) tall, round, erect, lower part tinted red. Raceme 7 in (18 cm) long, with 16 or more flowers in loose formation. Bracts emerge before flowers and wither, but do not fall at anthesis. Flowers tinted purple, becoming a deep, nearly red shade, the corolla always remaining closed. Anthers deep violet. Flowers mid-to-late summer.

Habitat: In woods in the province of Keiki, Korea. In Korea it is a common but much-loved garden plant.

Illustr.: None available.

Var. *normalis.* Rhizome very long below ground with basal shoots bunched up tightly. Blade slightly smaller than type, scape considerably taller, up to 26 in (62 cm), round, rigid, and dark purple at base, sometimes with one or two foliaceous bracts. Raceme upright in spike form. Flower bracts ovate, persisting after anthesis. Flowers blue-purple, funnel-shaped and open, unlike type.

Illustr.: AHS Bulletin, No 3, p. 33. *Hosta: The Flowering Foliage Plant (jacket).*

(e) SECTION PICNOLEPIS

H. longipes	(Franchet et Savatier) Matsumura (1894)
Funkia longipes	Franchet et Savatier (1876)
H. sieboldiana var. *longipes*	Matsumura (1905) Japanese name: *Iwagiboshi* (The Rock Hosta) by T. Iwasaki (1916)

The Japanese name alludes to this plant's wild habitat. It grows in the crevices of rocks, on shallow cliff faces and in moss on tree trunks, in places of high humidity, river valleys or in the spray of waterfalls, often in full sun. Its specific name (which means with long feet) refers to the way it pushes its roots deep into crevices. It also grows in shallow soil at the edge of mountain streams and is sometimes entirely submerged with water after heavy rain. It has a wide area of distribution and is, therefore, extremely variable and hybridizes freely with *H. kikutii* (p. 80). In spite of its habitat it is extremely drought-tolerant.

There is a taxonomic problem with this species in that the type specimen of Franchet et Savatier does not agree with the description published by Maekawa whose diagnosis follows. There are many botanical varieties and forms, some of which have not yet received a Latin diagnosis. A few of these are now grown in Britain, usually from seed. *H. longipes* is an unusual hosta on account of its habit and place of growth in the wild and it deserves to be better known as a garden plant in the west, although it is very popular in Japan where it is often grown as a pot 'bonsai' plant. Elsewhere it is still a collector's plant.

*

A medium-sized hosta, making a vigorous clump. Rhizome compact, roots long, far-searching. Petioles rigid, broadly channelled, dotted dark purple, winged in upper half, $6\frac{1}{2}$ in (16 cm) long. Blade cordate-ovate, or elliptic-ovate, occasionally orbicular, apex acuminate: upper surface dull, mid-to-dark green, slightly undulate margin, $5\frac{1}{2}$ in (14 cm) long × 3–4 in (8–10 cm) wide, with 7–8 pairs of veins; glossy below, veins glaucous, the base of the midrib dotted purple on both sides. Scape short, round with dark purple dotting down its entire length, obliquely poised, horizontally or even vertically from the leaf-mound depending on habitat. Bracts whitish-purple, softening and withering before anthesis. Flowers bell-shaped, wide-tubed, pale mauve to milky-white, dark purple on reverse. Flowering from late summer into autumn.

Habitat: Central Honshu and Kyushu.

Illustr.: Gentes Herbarum, 2, 138, 1930, *Jour. Fac. Sci. Tokyo,* Botany, 5, 323, 1940.

H. longipes has, in recent years, been found in one area of Mt Fuji in company with *H. kiyosumiensis, H. montana* and a green, lanceolate-leafed hosta which is likely to be *H. sieboldii.* (The Japanese refer to all small, lanceolate, all-green-leafed hostas with tall scapes as *Koba-Giboshi.*)

Forma *hypoglauca* is very rare in the wild. It has reddish stippled petioles, leaf blade with glaucous dusting, blue above, white below. This dusting can be rubbed off, unlike that on *H. hypoleuca* which is permanent. Flowers white; blooms earlier than type.

Forma *viridipes* (Maekawa 1940) is the rarest of all the forms of this species in the wild. Petioles bright green, leaves lanceolate, purple dotting on scape, near-white flowers.

Var. *latifolia* (Maekawa 1940) is distinguished by its broad leaves, distinctly purple at base. (It is often mistaken for *H. rupifraga.*) A form of var. *latifolia* exists which has wide, shining golden leaves in spring; it is known in Japan as 'Amagi-nishiki'. There is no Latin description for it yet.

Var. *caduca* (Honda) blooms in October. By some it is still today thought to be *H. tardiflora,* especially in Japan. Others think it may be *H. sparsa.* It is prized in Japan as a late-flowering pot plant.

Var. *aequinoctiiantha* is so-named because it supposedly flowers at the autumnal equinox.

H. rupifraga   Nakai (1930)
Japanese name: *Hachijo-giboshi*

This is one of the most outstanding hosta species – a plant of great garden metit. It has thick, leathery, shiny, heart-shaped, undulate leaves, and forms well-furnished mounds. It produces dense racemes of deep mauve flowers in September and October, at which season it associates well with *Liriope platyphylla* cultivars. An exhibit of the flower scapes of *H.*

rupifraga sent by the Royal Botanic Gardens, Kew, to the Royal Horticultural Society's Great Autumn Show, was given an Award of Merit.

H. rupifraga was introduced to Britain by Elliott Hodgkin in the 1960s and although now grown in one or two botanic gardens, it is still scarcely known. In the wild it grows on rough, rocky mountainsides. Its name means 'rock-breaker' from the way its roots penetrate rock fissures. It is sometimes confused with *H. longipes* var. *latifolia* (p. 86).

*

A medium-sized hosta forming a dense clump. Rhizome short, thick. Petiole $1\frac{1}{2}$–$2\frac{1}{2}$ in ($3\frac{1}{2}$–6 cm) long, shallowly channelled, winged on upper part, green. Blade firm, leathery, broadly ovate-cordate, slightly concave, of thick substance, funnel-shaped at base; up to $4\frac{3}{4}$ in (12 cm) long × $2\frac{3}{4}$ in (7 cm) wide; dark yellowish-green; upper surface smooth and slightly lustrous, underside pale green, with 6–8 pairs of veins, impressed above, shiny and elevated below. Scape stout, up to $15\frac{1}{2}$ in (40 cm), bearing short, dense racemes. Flower bracts navicular, obliquely spreading during anthesis, firm and pale purple. Flowers round, bell-shaped, not opening wide, lustrous pale violet. Anthers purple. Capsules short and thick. Flowering from early autumn.
Habitat: Only found on Hachijo Island.
Illustr.: Jour. Fac. Sci. Tokyo, Botany, 5, 390, 1940; *Hosta: The Flowering Foliage Plant,* Fig. 23.

H. tardiflora   (Irving) Stearn 1965
Funkia tardiflora   Irving
H. lancifolia var.   Bailey (1930)
   tardiflora
H. sparsa   Nakai (1930)
Japanese name: *Aki-giboshi*

Introduced to Europe by Max Leichtlin about 1895 (p. 45), it is thought by some botanists to be conspecific with *H. sparsa* Nakai but with clonal variations. It may turn out to be a cultivar. An important hosta as it is the pod parent of the Tardiana group. It is late flowering (September and October), and associates well with mauve asters, liriopes and colchicums. Needs protection in winter in the coldest parts of Britain. Its thick, leathery-textured leaves render it almost impervious to slug and snail and cold-water damage. A large number of the flowers open together and remain open in good condition for many days, which makes it an excellent garden plant. It is particularly suitable for edging purposes.

*

The creeping rhizome is thick and dense with white horizontal fibres. Petiole longer than leaf-blade, up to $10\frac{1}{2}$ in (27 cm); erect, very green, dull purple-tinted at the base; broadly channelled and not winged. Blade erect, up to 6 in (15 cm) long × $2\frac{1}{2}$ in (6.5 cm) wide, narrowed at base, not decurrent, acuminate, lanceolate to narrowly elliptic; margin flat or shallowly undulate; thick and firm in texture; dark olive-green, shiny above, dull below with 5 pairs of veins, elevated on underside. Scape held slightly above leaf-mound, up to 13 in (35 cm), ascending at 45°; raceme of 25 or more flowers, dense at first, elongating and becoming less dense during flowering. Flower bracts narrowly ovate, white to purplish, withering after anthesis. Narrowly funnel-shaped flowers, bright mid-violet; anthers purple. Flowers autumn.
Habitat: Gardens of Japan. No wild location known.
Illustr.: *Acta Horti Bergiani,*m 16, pl. 20, 1954; *Hosta: The Flowering Foliage Plant*, Fig. 11.

H. sparsa   Nakai (1930)
Funkia japonica var.   *Kew Handlist* (1902)
    tardiflora   Wright (1916)
                       Japanese name:
                           Akigiboshi

Virtually identical with *H. tardiflora* and considered by some botanists to be conspecific, exhibiting only clonal differences. Not found in the wild. Collected by Mr Matsuzaki from gardens in Nagoya and in cultivation in Tokyo.
Illustr.: None available.

H. hypoleuca   Murata (1962)
                       Japanese name:
                           Urajiro-giboshi

The final sectional placement of *H. hypoleuca* is still in some doubt and it may only temporarily fit best into Picnolepis.

H. hypoleuca is one of the loveliest of all hostas, though still little-known in Britain and Europe. Its Japanese name *Urajiro* means 'white-backed', which aptly describes this hosta, for an intense, nearly white, glaucous pruina covers the under-side of the leaf. In the wild *H. hypoleuca* grows on vertical south-facing cliffs, mountainsides or rock faces and the white pruinosity beneath the leaves is there to protect them from the reflected heat of the rocks. As it matures, the number of leaves decreases and the size of the remaining leaves increases. The roots find nourishment in crevices of the rock faces, but an additional source of nutriment is the falling debris and detritus which collects in the crown, gradually forming a mat of humus which protects and sustains the roots. *H. hypoleuca* is also found on the wet side of the mountains where its habit of growth is similar to that of *H. longipes*. In cultivation in rich, well-manured soil, it forms a clump of luxuriantly large leaves. The leaves are a very pale soft green covered in a thin glaucous bloom which disappears if the hosta is grown in full sun and the leaf surface becomes very shiny indeed. The pale lavender to milky-white flowers extend on almost horizontal scapes, nearly touching the ground. The blooming season is very short.

*

Medium-sized to large hosta, forming open clumps in cultivation. Rhizomes compact with long, searching roots. Petiole deeply channelled, up to 14 in (34 cm), pale green, glaucous, dotted purple. Blade up to 18 in (45 cm) × 12 in (30 cm) wide; broadly ovate, very slightly undulate, cordate at base; sometimes decurrent; margin slightly rippled, soft, of leathery texture, medium substance, shiny above (in full sun, otherwise glaucous), dense mealy-white below, with 8 pairs of widely spaced veins. Scape up to 14 in (35 cm), not exceeding leaf-mound, almost horizontally angled depending on habitat, glaucous, bearing raceme of 10–18 pale mauve to milky-white, bell-shaped flowers. Occasional large foliaceous bracts near top of scape. Flower bracts ovate to oblong, very pale in colour; anthers yellow to purple. Flowers midsummer for a short time only.

Habitat: Aichi and Shizuoka Prefectures.

Illustr.: Acta Phototax. 19.67, 1962; *AHS Bulletin,* No. 4, p.11; *Hosta: The Flowering Foliage Plant,* Fig. 15.

H. tortifrons   Maekawa (1940)
Japanese name:
Kogarashi-giboshi

An unusual rock garden hosta which is known for its contorted and twisted narrow leaves. New to British gardens, but has been known in America for a number of years.

*

In some respects it is similar to *H. tardiflora,* i.e. in flowering time, leaf substance. Makes an upright, open mound. Blade decurrent, twisted and contorted down the entire length, including the margins; of thick substance and leathery texture. Dense raceme of purple flowers in autumn.

Habitat: Cultivated in gardens.

Illustr.: AHS Bulletin, No. 3, p.38.

H. okamotoi   Araki (1942)
Japanese name:
Okuyama-giboshi

Little known outside Japan and not cultivated in Britain. Related to *H. rupifraga* (Nakai) but

easily distinguished by the sparse inflorescence and different-shaped tube. The fertile flowers are white and funnel-shaped. Blooms during September and fruits in November. Found growing on tree trunks on shady mountain slopes in Tamba by Mr Shogo Okamoto after whom it was named.

Habitat: Honshu, Kyoto Prefecture, Kita-kuwa-dagun, Chiimura.

Illustr.: None available.

H. takiensis   Araki (1942)
Japanese name:
Takigiboshi

Collected by Y. Araki in Kyoto Prefecture and, erroneously allied by him to *H. lancifolia.* Little known outside Japan and not cultivated in Britain.

*

Short rhizome. Grooved petioles $3\frac{1}{4}$–$4\frac{3}{4}$ in (8–12 cm) long, spotted dull purple. Leaf blade widely ovate, shiny, leathery and thick in texture, $4\frac{3}{4}$ in (11 cm) long × $1\frac{1}{4}$ in (3.5 cm) wide. 6–8 pairs of veins, not raised. Scapes erect, spreading, up to 13 in (32 cm) tall, purple-dotted towards the top; 2–4 foliaceous bracts. Flower bracts very wide and spreading, green suffused white. The funnel-shaped near-white flowers have purple-dotted yellow anthers; fertile. Flowers late summer and autumn. Fruits oblong. Long, large seeds.

Illustr.: None available.

(f) SECTION TARDANTHAE

H. tardiva   Nakai (1930) *Bot. Mag. Tokyo*
Japanese name:
Nankai-giboshi

H. tardiva is the type species for Section Tardanthae. Sometimes mistaken for *H. lancifolia* which it slightly resembles, *H. tardiva* flowers a little later and has wider leaf blades. Both are sterile. It is also similar to *H. cathayana* (of the same section) which is a smaller plant. Much cultivated in Japan since the beginning of the current craze there for wild plants.

*

Rhizomatous root stock. Petiole grooved and winged, lower part dotted dull purple; up to 3 times as long as blade. Many erectly poised leaves spread out from the rootstock. Blade ovate or ovate-oblong, base tapering or blunt, margin shallowly undulating, apex acuminate; up to 5 in (12.5 cm) long × 1 in (5 cm) wide; bright, shining dark green above, paler underneath, with 8 pairs of veins. Scape at least twice as long as blade, spotted dark purple in lower part, up to 15 in (37 cm) tall. Raceme loose with few, large flowers. Flower buds similar in colour to those of *H. clausa*, bright mauve on opening, disposed on one side of scape only. Anthers dull bluish-violet. Flowers from early autumn onwards.

Habitat: Shikoku Island, Kii peninsula.

Illustr.: *Jour. Fac. Sci. Tokyo*, (Botany, 5, 1940, pl. 71–72.

A form with striated variegation has been called *H. tardiva* 'Variegated' but this has, as yet, no Latin diagnosis and may not even be the same species.

H. okamii

(Maekawa) Maekawa (1940)
Japanese name:
Murasame-giboshi

This hosta is, according to Maekawa, related to *H. tardiva* and has similarities in the flower with *H. lancifolia*, but he has, nevertheless placed it in a different section.

It is probably not cultivated outside Japan.

*

Smallish hosta, with stiff spreading leaves. Rhizomes very short, with stoloniferous roots. Petioles broadly winged, shorter than blade, up to 2¾ in (7 cm) long. Blade lanceolate to oblong-lanceolate, apex narrowly acuminate; wedge-shaped and decurrent at base; up to 4 in (10 cm) long × ½ in (2 cm) wide; margin slightly undulate; intense green and slightly lustrous, with 4 pairs of veins. Scape round, slender, up to 20 in (50 cm) tall. Raceme lax, approximately 12-flowered. Flower bract thinly membranaceous, navicular and translucent at anthesis; nearly as long or longer than pedicel. Flowers funnel-shaped, pale purple; stamens far surpassing corolla. Anthers whitish with pale purple margins and tips. Flowers very late. Autumn.

Habitat: Honshu.

Illustr.: None available.

H. gracillima

Maekawa (1936)
Japanese name:
Hime-Iwa-giboshi
(named by O. Okomoto in 1934)

The Japanese name alludes to its habit and its size. It is found growing in rocky mountainous places, on cliffs and in valleys. It has a very limited distribution. Being one of the smallest hosta species, it is very suitable for use as a rock garden or peat bed plant. The name 'gracillima' has been misapplied to many hostas. It is easy to cultivate and grows very well as a pot plant, for which purpose it is often used in Japan. It is also a perfect specimen for a sink garden.

*

Rhizome short, leaves spreading. Petiole up to 2¼ in (6 cm) long, usually much less, very slender and deeply grooved. Blade lanceolate, margin undulate, apex acuminate, blunt at base, ¾–2¾ in (2–6 cm) long × ½–1 in (1–2 cm) wide. Upper surface matt or shiny (collectors have recorded both), slightly shiny underneath, with 3 pairs of veins. Slightly decurrent. Scapes erect or slightly angled, not ridged, up to 9½ cm (25 cm)

tall. Bracts linear, spreading. About 10 flowers per raceme, loosely arranged and disposed to one side of the scape. Flowers widely funnel-shaped, pinkish-violet. Anthers blue. Capsule small, cylindrical. Flowers in autumn.
Habitat: Shikoku Island.
Illustr.: Jour. Fac. Sci. Tokyo, Botany, 5, pl. 75–6.

H. pulchella
Fujita (1942)
Japanese name:
Ubatake-giboshi

Pulchella in Latin means beautiful or little and this hosta is both. A dwarf plant not yet known in Britain or Europe though plants have reached the United States from Fujita himself. It is much sought-after in Japan on account of its relatively large flowers, sometimes described as fragrant. The leaf blade is little bigger than that of *H. venusta.* A useful addition to the small number of hostas suitable for the rock garden or peat bed. Fujita mentions that there is a great variation in the wild in this species.

*

Dwarf hosta of stoloniferous habit, forming small, neat clumps with semi-erect leaves. Rhizome creeping, narrow. Petiole $\frac{3}{4}$ in (2 cm) long, sparsely dotted deep purple. Blade ovate to cordate, apex gradually acuminate; margin flat; up to 2 in (5 cm) long × $\frac{1}{2}$ in (1.5 cm) wide; upper surface vivid glossy green, with up to 4 pairs of veins, elevated and glabrous below. Scape obliquely ascending, well over-topping leaf mound. Foliaceous bracts lanceolate. Flower bracts greenish, navicular, closed before and at anthesis. Flowers campanulate, more or less fragrant (this characteristic not noted in American plants), pale lavender, 1–15 per raceme. Stamens protrude from perianth. Anthers oblong, purple before opening. Fertile. Capsules cylindrical. Seeds small. Short flowering season, late mid-summer.
Habitat: 1600 ft altitude on Mt Sobo, on outcrops and ridges, Ooita Prefecture, Kyushu.
Illustr.: None available.

A glaucous-leafed form is known to exist locally called 'Urajiro', but there is, as yet, no published description.

A yellow-margined form has recently been found in the wild but is not especially attractive, although it is interesting on account of its rarity.

H. tibai
Maekawa (1984)
(this spelling used in the scientific literature)
H. chibai (spelling used by Kanecko & Ohwi)
Japanese name:
Nagasaki-giboshi

This is the Nagasaki hosta, so called because it grows wild in the Japanese cedar (*Cryptomeria*) forests and forest margins and valleys circling the town of Nagasaki. It is almost certain that Siebold brought it back to Europe but no recorded proof exists. It is not grown in Britain or Europe although plants have been growing in America since 1971. The description I give below is taken from notes sent to me by George Schmid, derived from the Latin diagnosis of Maekawa (1984) and from George Schmid's observations of his own plants.

*

Small to medium-sized. Blade erect-spreading; ovate to ovate-elliptic; $6\frac{1}{2}$ in (15 cm) × 4 in (10 cm) wide at the most; apex abruptly acuminate; base cordate and distinctly petiolate; upper surface shiny, mid-green, with 4–7 pairs of veins. Scape upright, rising well above leaf-mound and well-branched (Maekawa does not mention this); up to 20 in (50 cm); raceme generally 4 in (10 cm) long; fairly dense; flowers disposed to one side of scape. Flower bracts green, lanceolate, navicular. Flowers horizontal before anthesis, bell-shaped, light purple, stamens protruding from perianth. Anthers pale yellow, dotted purple. Blooms very late – short flowering season, late September to early October.

Habitat: Nagasaki Prefecture, Kyushu.
Illustr.: None available.

H. takahashii — Araki (1942)

Japanese name:
Schichizo-giboshi

According to Araki this hosta is generally similar to *H. tardiva*, but it has a different local name. It differs from *H. tardiva* in its widely ovate leaf blade. It was collected by Schichizo Takahashi. Araki, who grew it in his own garden, named it in Takahashi's honour and published a description of it. Fujita considers it to be synonymous with *H. tardiva*. It is not known in Britain or Europe.

*

Leaf blades widely ovate. Flowers thin-textured, funnel-shaped, pale purple. Anthers purple and glaucous. Fruits oblong, seeds medium-sized. Blooms late summer to early autumn.
Habitat: Mountains of Ibukiyama, Honshu.
Illustr.: None available.

H. tsushimensis — Fujita (1960)

Japanese name:
Tsushima-giboshi

This hosta, very common on Tsushima Island, is also found on the mainland in Nagasaki Prefecture. The type was found on Mt Ariake, Nagasaki Prefecture, and several other locations outside Tsushima are on record (Schmid, personal communication). Unlike most hostas it grows in the wild in rather dry conditions instead of the damp areas normally associated with hostas, but it is found in a wide variety of habitats. It is known in Britain only from seed-raised plants. My own plant is much smaller than Fujita's description.

*

Small to medium-sized. Rhizome short, erect. Leaves semi-erect, tufted. Petioles up to 9½ in (25 cm), dotted dark purple. Blade ovate to ovate-cordate, up to 8½ in (22 cm) long × 4½ in (11 cm) wide, apex broadly acuminate; upper surface mid-green, slightly lustrous, with 5–9 pairs of veins, elevated and glabrous below. Scapes obliquely ascending, well above leaf-mound; up to 40 in (75 cm); purple-dotted. Foliaceous bracts obliquely ascending, dotted purple. Flower bracts green, navicular, never fully open. Flowers funnel-shaped, greenish in bud, almost white, but variable with protruding stamens. Anthers purple-dotted. Capsules cylindrical, seed medium-size. Blooms late summer to early autumn.
Habitat: throughout Tsushima and Nagasaki Prefecture.
Illustr.: None available.

Forms with pure white flowers exist as does a tri-colour form, dark crimson on the base of the tube, purple petals and white round the transparent lines.

H. cathayana — Nakai ex Maekawa (1940)

H. japonica var.
fortis — Bailey (*Gentes Herbarum* 1932)

Japanese name:
Akikaze-giboshi

Very similar to *H. tardiva* but with leaves smaller and more glossy or lustrous, and the flower bracts a brilliant green. Scape slender, round, up to 30 in (75 cm), sometimes bearing 2 or 3 foliaceous bracts. Flowers lavender. Flowering in late summer.
Habitat: Western Honshu, Japan, Nanking, China.
Illustr.: None available.

(g) SECTION NIPPONOSTA

(i) Sub-section Nipponosta-vera

H. longissima — Honda (1935)
Funkia lancifolia angustifolia — Regel (1876)

Funkia Mizu-
   giboshi, Sasi-
   giboshi
H. japonica var.
   lancifolia

Y. Iinuma, Somoku
   Dzuetsu (1874 or
   1876)
Hondsa (1930)
Japanese name:
   Mizu-giboshi,
   Nagaba-giboshi

H. longissima is one of the few species where
both the herbarium specimens and the botanists'
descriptions agree with the present-day living
material. The only point of disagreement is that
Maekawa's var. *brevifolia* (1940) appears to be
conspecific with the type plant, an identification
borne out by the fact that Maekawa gives var.
brevifolia the names Sasi-giboshi and Mizu-
giboshi (Iinuma). The same name, though
spelled 'Sasi-Kiboshi', is written in Siebold's
hand beside one of his Japanese specimens of
Hemerocallis lanceolata in his herbarium at
Leiden, and this specimen exactly corresponds
both with the woodcut and description given
by Iinuma in Somoku Dzuetsu (1974), and to
living material in cultivation in Europe today.

Siebold himself later changed the name to
Funkia lanceo-lata and this is how it is described
in his nursery catalogue of 1844. This name was
still being used for *H. longissima* until at least
the 1890s by nurseries in America, suggesting
that the American plants were imported from
Europe rather than Japan.

H. longissima is quite distinct from all other
hostas on account of its narrow, grass-like leaves
and weak-growing reddish flowers. It flourishes
in very damp conditions and is highly susceptible to drought. In the wild it grows in marshy
places and open grasslands where it merges so
well with the grasses that it is quite difficult to
spot unless in flower.

Many different forms abound in the wild.
Two recent finds are *H. longissima* 'Aureo
Nebulosa', the cloudy variegation not fading to
green, and a white-flowered form. Two quite
distinct forms have recently reached British
gardens. One with leaves $\frac{3}{4}$ in wide and up to
10 in long, with an indented variable centred

Fig. 10 *H. tokudama* var.
aureonebulosa
Description on p. 77

variegation, one with leaves $\frac{3}{4}$ in wide and 3 in long, markedly recurved at the tip. Two variegated cultivars *H. 'Asahi Comet'* and *H. 'Asahi Sunray'* have recently reached America and have been registered as cultivar names. However, according to Japanese sources this species is providing a wealth of white-margined and yellow-margined and streaked forms which have yet to be described and given scientific or cultivar names.

Most of the hostas being grown in Britain under the name *H. longissima* are in fact a narrow-leafed form of *H. sieboldii*. These may easily be distinguished from *H. longissima* by their yellow anthers and the markings which resemble white clasped hands on the lobes.

*

Small to medium-sized, mound-forming hosta, with leaves upright to arching. Rhizome long, creeping. Petiole shallowly channelled, narrowly winged, without spotting, up to $9\frac{1}{2}$ in (24 cm) on mature plant, growing in damp conditions. Blade linear-oblong, arching, base blunt; narrowly decurrent; apex gradually acuminate; 7 in (17 cm) long × $\frac{3}{4}$ in (2 cm) wide; texture slightly leathery, upper surface matt, mid-to-dark green, slightly shiny below with 3–4 pairs of veins. Scapes numerous, compensating for the sparsely flowered raceme, up to 22 in (55 cm), held well above leaf mound. Flower bracts shiny, green, navicular. Flowers funnel-shaped, pale reddish-mauve with purple veins. Anthers purple-blue. It is an extremely variable hosta.

In the form the Japanese call *Nagaba-mizu* the leaf is much longer than the type.
Habitat: Originally thought to be west of Aichi Prefecture, Honshu, but a new community has been found in Hamanatsu in Shizuoka Prefecture, north of Aichi.
Illustr.: Acta Horti Bergiani, 16, pl: 21, 1954. Aden, *The Hosta Book,* 1988.

H. helonioides

Maekawa (1937)
Japanese name:
Hakama-giboshi

Found by Dr A. Kikuchi in Akita Prefecture, Honshu. Gets its specific name from a supposed resemblance to *Helonias* and, indeed, the leaf shape and general aspect of the two plants have much in common, though the hosta has distinct petioles which *Helonias* does not. It is generally only grown in its variegated form 'Albo picta', but since the flowers are showy and presented well above the foliage, the typical plant may have more of a future as a garden plant. It has an unusually wide petiole. It is a vigorous hosta which will tolerate a great deal of sun. In Japan it is much used as a pot plant, particularly in the variegated form.

*

Small hosta, forming spreading clumps. Rhizome slightly stoloniferous. Petiole short, up to $6\frac{1}{2}$ in (16 cm), shallowly grooved, broadly winged, sparsely purple-dotted at base. Blades linear-elliptic, apex rounded; margin flat or slightly undulate; up to 8 in (20 cm) long × 1 in (2.5 cm) wide; strongly decurrent; dull, mid-olive-green above, shiny below, with 3–4 pairs of veins; thin substance. Scapes slender, erect, purple-dotted at base, well over-topping leaf-mound, up to 24 in (60 cm), with navicular bracts just below raceme. Flower bracts green, oblong, lanceolate, persisting after anthesis. Raceme lax, bearing 10–20 campanulate, reflexed, mauve, purple-streaked flowers, 2 in (5 cm) long, dependent. Anthers small, pale blue. Flowers in late summer.
Habitat: Akita Prefecture, Honshu.
Illustr.: None available.

Forma *albopicta* is a striking pale cream-margined form. An attractive garden plant.
Illustr.: Jour. Fac. Sci. Tokyo, Botany, 5, 40; *Hosta: The Flowering Foliage Plant,* Fig. 17.

H. rohdeifolia

Maekawa (1937)
Japanese name:
Omoto-giboshi

So named on account of its resemblance to typical plants of *Rohdea japonica*, a genus in *Convallariaceae* named in honour of Michael Rohde, a physician and botanist of Bremen. The typical *H. rohdeifolia* is a variegated plant, having the leaves pale yellow-margined. It is known only as a cultivated plant, found growing in gardens in Chugoku distrct, western Japan, although a green-leafed form, forma *viridis* (Maekawa 1940) was found in the wild by Professor Kikuchi on Mt Hiei, Kyoto Prefecture. *H. rohdeifolia* is little known in Britain and Europe but is popular in Japan, both as a garden plant and as a pot plant.

*

Medium-sized hosta. Rhizome slightly stoloniferous. Petiole yellow-margined, longer than leaf blade, up to 15 in (38 cm). Blade held erect: lanceolate to oblong-lanceolate; margins slightly involute; apex acuminate; narrow at base; up to 7 in (18 cm) long × up to 2½ in (6 cm) wide; upper surface lustrous, deep olive-green, with 5–6 pairs of veins. Scape extends high above leaf-mound, up to 40 in (101 cm) tall, erect, bearing several bright green, creamy-white margined foliaceous bracts near the centre. Flower bracts ovate; navicular, whitish-yellow-margined, persisting after anthesis. Flowers nodding, pale purple, striped deeper purple. Anthers oblong, small, creamy-white. Flowers at midsummer.
Illustr.: None available.

Forma *viridis* (Maekawa 1940) has markedly green leaves.

H. ibukiensis   Araki (1942)
Japanese name:
Ibuki-giboshi

Related to *H. sieboldii*, but differing in its dark green leaf blade, in the shape of the perianth tube, and in being only marginally fertile, and more usually sterile. Flowers funnel-shaped, purple; stamens yellow, shorter than perianth. Anthers yellowish-white. Fruits oblong to lan-

ceolate. Seeds small. Flowers in late summer. Fruits in October.
Habitat: Mt Ibukiyama, Shiga Prefecture, Honshu.
Illustr.: None available.

H. lancifolia   (Thunberg) Engler 1888

Aletris japonica	Thunberg (1780)
Hemerocallis japonica	Thunberg (1784)
Hemerocallis lancifolia	Thunberg (1794)
Funkia lancifolia	Sprengel (1825)
H. lancifolia var. *thunbergii*	Stearn 1932
	Japanese name: *Sazi-giboshi*

The first hosta ever described in a European publication and also the first to receive a scientific name. Its historical importance is fully covered on p. 33. Imported into Holland by Siebold in 1829, material was sent to Britain very soon after, since Loddiges described the plant in his *Botanical Cabinet* in 1831. In spite of this it is probable that *H. lancifolia* is not a true species at all: certainly it is not fertile. Some authors have suggested that *H. lancifolia* and *H. cathayana* are the same plant, but they are morphologically quite distinct.

Maekawa's incorrect use of Stearn's epithet *H. lancifolia* var. *thunbergii* is fully discussed on p. 49.

H. lancifolia is well-established in cultivation in Britain, Europe and the United States. It is a slender, elegant garden plant producing an abundance of good purple flowers on tall, graceful scapes in late summer. It seldom sets seed. It makes excellent ground cover.

*

Medium-sized, loosely clump-forming. Rhizome becoming stoloniferous when cultivated in soft ground. Petiole slender, grooved and open-winged, red-dotted at base; up to 13 in (31 cm) tall. Blade decurrent to petiole, lanceolate or oblong-lanceolate, margin flat, narrowly acuminate; 4–7 in (10–17 cm) long × 1 in (2.5 cm) wide; thin substance, glossy, dark olive-green, slightly shiny below, with 5–6 pairs of veins. Scape slender, much exceeding leaf-mound, up to 18 in (45 cm) tall, characterized by several large foliaceous bracts up to 2½ in (6.5 cm) long which closely resemble the basal leaves; base red-purple-dotted. Flower bracts shiny green, narrowly navicular before anthesis. Raceme erect with up to 30 well-spaced, narrow, tubular, strong purple flowers. Anthers violet mauve.

Habitat: Of garden origin.

Illustr.: *Acta Horti Bergiani*, 1954, fig. 2, 3 and pl. 17; *Hosta: The Flowering Foliage Plant*, Fig. 6.

H. lancifolia 'Aurea' (hort.). Larger than the type. 4–5 pairs of veins. It is known in two forms, one viridescent, that is, losing its yellow colour by midsummer, the other remaining yellow all the summer. The second is very rarely seen in gardens but is often confused with *H.* 'Subcrocea'.

The cultivar *H.* 'Sea Sprite' once thought to be a superior form of *H.* 'Kabitan' is more likely to be related to H. *lancifolia*.

H. clavata	Maekawa (1938)
H. japonica var.	Makino (1910)
intermedia	Japanese name: *Ko-giboshi* or *Murasino-giboshi*

This hosta is not cultivated in Britain or Europe, and it is not even certain whether or not it should be given species status. My description of it is based on that of Maekawa (1940).

*

Blade lanceolate or oblong-lanceolate, narrowly decurrent to the petiole; margin slightly undulate, apex acute; up to 4 in (10 cm) long × 1½ in (4 cm) wide; dull green with 3–4 pairs of veins. Scape slender, erect, up to 20 in (50 cm). Flower bracts acute, green, often without colour at anthesis. Pedicels curved, very short. Bud bluntly club-shaped. Flowers funnel-shaped, white suffused purple. Anthers yellow with bluish margins.

Habitat: Saitama, Kanagawa and Tokyo Prefecture; in mountainous areas of central Honshu.

Illustr.: None available.

H. alismifolia	Maekawa (1984), *Jour. Jap. Bot.* Japanese name: *Baran-giboshi*

Named for its fancied resemblance to *Alisma* – a water plant. A name of uncertain application, for while Maekawa describes a plant with rather erect leaves, with an overall length (including the petioles) of 3 ft (90 cm) later photographs from Japan show a plant more like *H. sieboldii*. There is also the question as to whether the plant described by Fujita and placed by him in Section Stoloniferae is the same as that described by Maekawa.

Provisionally placed in this section because it is a Japanese hosta and its identity is too unclear to justify placing it elsewhere.

Illustr.: None available.

H. calliantha	Araki (1942) Japanese name: *Fuji-giboshi*

Calliantha means 'having beautiful flowers'. Similar to *H. rectifolia* from which it differs in some details. Blooms late summer to early autumn. Anthers yellowish-white. Flowers purple, funnel-shaped and well-opened. Triangular fruits.

Habitat: Honshu, Kyoto and Hyogo Prefectures, Minami-kuwadagun, Higashi-betsuin-mura. Grows in damp open places at the foot of mountains. Not known in Britain.

Illustr.: None available.

H. campanulata	Araki (1942) Japanese name: *Tsurigane-giboshi*

Araki's holotype of this hosta has been studied in the United States and findings there show that it is similar to *H. sieboldii* (Paxton), although Araki himself says that the similarity is with *H. rectifolia* (Nakai) (in some respects) although it exhibits major differences, i.e. shape of perianth tube, anther colour (yellowish–white in *H. campanulata*) and the smaller leaves. It is not known in Britain.

*

Rhizome produces thick, long shoots. Petiole winged, faintly purple-dotted. Foliaceous bracts on scape. Blooms mid-autumn. Very fertile. Fruits triangular.
Habitat: Honshu, Kyoto and Hyogo Prefectures, Takigun, Kumobemura. In damp places at the foot of mountains.
Illustr.: None available.

H. sieboldii

(Paxton) Ingram (1938)

Hemerocallis sieboldii
H. albomarginata (Hook.) Hyl. Ohwi
H. lancifolia var. (Thunberg) Engler
   albomarginata
H. sieboldiana misapplied

Japanese name: *Koba-giboshi*

One of the commonest hostas in the wild where it has a wide distribution. It was collected first in its variegated form which is therefore the type. *H. sieboldii* could also have been considered as the type plant for the subgenus *Giboshi* on account of its wide distribution and its polymorphic qualities. Unlike most species its habitat is open grasslands and fully exposed to sunshine. In common with the majority of other hostas it needs moisture at its roots. At anthesis, only the scape is visible as the leaf-mounds are surrounded by grasses and weeds as with *H. longissima*. The Japanese still confuse it with *H. lancifolia* (Engler) which is morphologically and taxonomically entirely different (p. 94).

There is much confusion over the identity of the many botanical variants and cultivars since the Japanese have been despatching to collectors all over the world plants and seeds bearing Latinized epithets, most of which are incorrect, applying to other hostas, or epithets which have never been validated. *H. sieboldii* is even confused in Japan with *H. longissima*, although from the material of both we have available in the West, it is hard to see why, except on casual inspection. *H. sieboldii* was originally brought to Europe by Siebold in 1830 under the name *F. spathulata folis albomarginatis*. It was described first by Paxton in 1838 (p. 42).

H. sieboldii is an excellent ground-cover plant, making dense clumps of white-margined or all-green spatulate leaves with flower scapes well-overtopping the leaf-mound. In cultivation *H. sieboldii* prefers light shade and will even tolerate dry places, but the leaves become yellowish in full sunlight and are susceptible to wind damage, in contrast to its preferred wild habitat (p. 42).

*

Small to medium clump-forming hosta. Rhizomes shortly creeping. Petioles shallowly grooved to almost flat, up to $10\frac{1}{2}$ in (26 cm), broadly winged, usually spotted purple. Blade lanceolate-elliptic, apex roundly acuminate, tapering at base; margin flat or slightly undulate, with narrow, clear white margins; matt, slightly rugose, dull mid-green upper surface, slightly, glossy below; 6 in (15 cm) × up to $2\frac{1}{2}$ in (6 cm) wide, with 3 pairs of veins but occasionally a fourth in mature plants. Scape slender, stiff, erect, rising to 20 in (50 cm), with one or two foliaceous bracts. Flower bracts very small, concave, intense green (slightly violet-tinged in bud), withering after anthesis. Flowers funnel-shaped, deep, clear mauve with deeper violet stripes and white lines. Anthers yellow. Blooms late summer to early autumn. Fruits large and heavy. Very fertile.
Habitat: Honshu, Shikoku, Kyushu, Southern Kuriles Island, Sakhalin Island.

Illustr.: Botanical Magazine 3657, 1838.
Thomas, G.S., *Plants For Ground Cover*, p. 29,
1970);
AHS Bulletin Vol. 2, p. 14 under *H. lancifolia
albo-marginato* (misapplied);
Hosta: The Flowering Foliage Plant, Fig. 16.

Var. *alba* (Irving) (Hyl.). Green-leafed, white-
flowered variant, sometimes incorrectly known
as *H. minor* 'Alba'.

Forma *kabitan*, erroneously placed in *H. lancifolia* by Maekawa; also erroneously known as
H. lancifolia 'Viridis-marginata'. Small, not
very vigorous hosta with stoloniferous habit
which, in Britain, often grows better in a pot
than when planted out. Blade lanceolate or
oblong-lanceolate, $4\frac{1}{2} \times 1\frac{1}{2}$ in (8.3 × 3.5 cm),
margins ruffled, conspicuously undulate (much
more than in the type), tapering to a point.
Very thin in texture, almost papery, surface
slightly shiny; bright, yellow-gold with narrow,
dark green margin. Leaves scorch if grown in
full sunlight. Slender scapes well above leaf-
mound. Flowers deep purple, in vivid contrast
to the foliage. Leaf colour fades very slightly at
anthesis. Much used as a pot plant in Japan. The
strange name *kabitan* is translated as follows:
ka = bright, brilliant; *bi* = beautiful: *tan*-red,
orange colour; the elixir of life.

Forma *subcrocea* has gold leaves.

Forma *shiro-kabitan* has small, green-margined,
white, elliptic-lanceolate, somewhat twisted
leaves. Used as a pot plant in Japan. Sometimes
incorrectly known as *haku chu-han*, Chapter 4,
(p. 114).

H. 'Yakushima-mizu', a dwarf form of *H.
sieboldii*. *Mizu* means water in Japanese. The
name was given by Y. Iinuma about 1910. It
does not in fact occur in Yakushima Island, but
comes from western Honshu, Shikoku and
Kyushu.

H. 'Shirobana kika'. Anthers appear out of the
flower bud before anthesis. Pure white flower
with variable number of petals (from four to

six). Very rare in the wild. The many hybrids
raised from *H. sieboldii* are listed in Chapter 4.

(ii) Sub-section Brachypodae

H. decorata	Bailey (1930)
H. 'Thomas Hogg'	misapplied
H. decorata f.	Stearn (1931)
marginata	Japanese name:
	Otafuku-giboshi

The historical background of *H. decorata* is
fully discussed on p. 29. It is one of a small
group of white-margined hostas about which
there has always been confusion, the others
being *H. undulata* var. *albomarginata* (syn *H.*
'Thomas Hogg'), *H. crispula* and *H. fortunei*
'Albo-marginata' (*H.f.* 'Marginato-alba' of gardens).

Furthermore, *H. decorata* has always been
known as 'Thomas Hogg' in America ever since
it was sent back to America by the nurseryman
and plant-collector of that name. It is also
one of the relatively few hostas of which the
variegated plant was described before the green-
leafed form was known. *H. decorata* is readily
distinguished from the other white-margined
hostas by its stoloniferous habit. It is sometimes
difficult to establish or to maintain – and,
consequently, it is rarely offered for sale in
Britain, but when well-suited on an acid, sandy
soil on the edge of woodland with plenty of
leaf-mould, it is a striking plant and will form
large colonies. It is much more vigorous in the
United States where it is much-used for ground
cover. The hotter summers in America may
also encourage vigorous growth. One of the
earliest hostas to sprout in spring.

The Japanese name *Otafuku* means 'round
face of a woman'. In times past it was believed
that *Otafuku* brings luck or happiness to the
home. But another meaning is ugly or 'not
beautiful woman'. *H. decorata* is not popular

in its native Japan in either the plain-green or the variegated form.

*

Small to medium-sized hosta, not clump-forming, stoloniferous, usually not vigorous. Rhizome, long, creeping. Petiole up to 12 in (30 cm) though usually less, broadly channelled, flattened out, broad-winged, margins flattish. Blade ovate-obtuse or nearly orbicular, shortly acuminate or rather blunt; base contracted, rounded; margins flat; $6\frac{1}{2}$ in (16 cm) long × $4\frac{3}{4}$ in (12 cm) wide; upper surface leathery, matt, dull, dark (sometimes very dark) green, with regular, distinct white margins, lustrous below with 4–6 pairs of veins; surface smooth, not impressed. Scape at least 20 in (50 cm) tall, overtopping leaf-mound by as much as three times. Foliaceous bracts dark olive-green, white-margined, clasping scape. Flower bracts dilated, broadly oblong-elliptic, yellow-margined. Raceme with up to 25 narrowly campanulate, dark violet flowers, fading after anthesis. Anthers yellow with good pollen. Capsules well-developed; seeds long and broad. Blooms midsummer. Very fertile.

Habitat: Honshu. Very little data available. Probably originally described from cultivated material.

Illustr.: Curtis's Botanical Magazine, 9395, 1935;
Jour. Fac. Sci. Tokyo, Botany, 5, pl. 89.

Forma *normalis* (Stearn), has plain green leaves, slightly rugose in mature plants, with apex more acute.

Forma *albi-flora* (only known in Japan) has wider leaves, ovate-orbicular and white flowers.

H. rectifolia   Nakai (1930)
H. longipes   misapplied
          Japanese name:
               Tachi-giboshi

So named on account of the semi-erect leaves.

The history of *H. rectifolia's* confusion of identity with *H. longipes* at Kew Gardens has been given on p. 47.

Related to *H. sieboldii, H. rectifolia* is one of the relatively few hosta species more remarkable for its flowers than its foliage. (Its flowers are similar to those of *H. decorata.*) It is very variable in the wild. Larger forms than the type grow in more northerly locations with scapes up to 34 in (100 cm).

Some variegated forms have recently been found. Of these, var. *chionea,* the first, has been validly published but no doubt others will follow. It is the commonest hosta in Hokkaido and Sakhalin Island.

H. rectifolia is now seldom offered for sale in British nurseries but well-established clumps are still growing in many botanic gardens and in the National Reference Collections.

*

Medium-sized to large, sturdy hosta, forming dense clumps with erect, ovate, elliptic leaves. Rhizome slightly soloniferous. Petiole shallowly grooved, up to 20 in (50 cm), broadly winged, no purple dotting. Blade ovate-elliptic, slightly involute, shortly acuminate, tapering at base, margin flat; 6–9 in (15–23 cm) long × $2\frac{1}{2}$ in (5 cm) wide; matt, dull, mid-green on upper and lower surface with 6–9 pairs of veins. Scape round, stout, erect, sometimes branched, occasionally dotted purple, 23–30 in (60–75 cm) tall; several conspicuous foliaceous bracts from the centre upwards. Flower bracts green, purple-striped; navicular, persisting after anthesis. Raceme begins low down on scape, many-flowered. Buds purple, flowers large, campanulate, violet with darker violet streaks inside tube, erect at first, finally nodding. Anthers narrowly oblong, whitish yellow with pale purple margins. Capsules well-developed, large, cylindrical, narrow. Flowers late summer.

Habitat: Central and northern Honshu, Hokkaido, southern Kuriles and Sakhalin. Found in damp mountain woods.

Illustr.: None available.

Forma *pruinosa* (Maekawa 1940) has the leaves, scapes and bracts covered with intense pruina.

Var. *sachalinensis* (Koidz.) Maekawa, is a geographical variant from Sakhalin.

Var. *chionea* (Maekawa 1940) has long-petioled, white-margined, lanceolate leaves, the autumnal leaves ovate-elliptic. It is often mistaken for *H. helonioides albo picta*.

H.r. var. *chionea* forma *albiflora* (Tatewaki 1934) has white flowers.
Illustr.: Jour. Fac. Sci. Tokyo, Botany, 5, 1940, pl. 94–95.

A round-leafed form, *Maruba-tachi-giboshi*, and a yellow-margined form, *Kinbuchi Tachi-giboshi*, have, as yet, no scientific descriptions.

H. opipara	Maekawa (1937)
	Japanese name:
	Nishiki-giboshi

H. opipara is an eye-catching garden plant and is also a welcome addition to those large, variegated-leafed hostas used for flower arrangement. It has a great future commercially. The whole plant is vigorous, has poise and distinction and is likely to rival *H. montana* 'Aureo Marginata' and *H. fluctuans* 'Variegated' in popularity. It has already been introduced to America but is not widely known in Britain yet. It was classified as a species by Maekawa in 1937 and validly described in 1940.

*

Generally similar to *H. rectifolia* and possibly a hybrid of it. It differs in the poise and shape of the leaves which are large, ovate-elliptic and horizontally spreading; margins slightly undulate, up to $\frac{2}{3}$ in (1 cm) wide, but variable, not regular. Petioles very broad. Blade $5\frac{1}{2}$–8 in (14–20.5 cm) long, 3–5 in (7.5–13 cm) wide; margin undulate, apex obtuse; with 9 pairs of veins, coriaceous, dark green, broad cream margin. Scape up to 24 in (60 cm), a little shorter than in *H. rectifolia*. Flowers lilac, bell-shaped,

Fig. 11 *H. tardiflora*
Description on p. 86

held horizontally. Anthers whitish-yellow, blue-margined.

Habitat: Reported from Aomori and Iwato Prefectures in north-east Honshu, but reports unconfirmed. Only known for certain in gardens. Maekawa based his species on a plant found by Dr A. Kikuchi, near Hirosaki City.

Illustr.: Jour. Fac. Sci. Tokyo, Botany, 5, pl. 96 and 97.

AHS Bulletin 1985.

Hosta: The Flowering Foliage Plant, (pl. 7).

H. atropurpurea   Nakai (1930)
                         Japanese name:
                              Kurobana-giboshi

This is reputedly a difficult hosta to cultivate, although I have seen a remarkably robust specimen growing at Wisley, the garden of the Royal Horticultural Society. It is said to have little vigour, and is reluctant to make leaf growth or to flower. Its leaves are said to die off in September even after coaxing in optimum conditions. It requires very moist soil. Professor Nakai described it from dried specimens. Now cultivated in America but still very little known.

*

Medium-sized hosta, forming dense clumps. Rhizome with short, creeping stolons. Petiole long, winged above centre. Blade lanceolate or oblong-lanceolate, up to 7 in (18 cm) long × 2½ in (5 cm) wide; matt, dark green, heavily pruinose on both surfaces, with 5–9 pairs of veins. Scapes up to 24 in (60 cm) tall. Flower bracts broadly ovate, purple, glaucous. Raceme lax. Funnel-shaped flowers, medium-sized, nodding, deep purple with glaucous bloom, curiously pinched towards the tip. Anthers yellow.

Forma *albiflora* (Tatewaki) has white flowers, though George Schmid suggests this plant is actually *H. rectifolia* var. *chionea* forma *albiflora* (personal communication).

Habitat: Mt Daisetzu, Hokkaido. In damp mountain valleys.

Illustr.: None available.

(h) SECTION FOLIOSAE

H. undulata	Bailey (1923)
Funkia undulata	Otto & Dietrich (1833)
Funkia ovata var. *undulata*	Miquel (1867)
Hemerocallis undulata	Siebold apud Miquel (1896)
	Japanese name: *Suji-giboshi*

One of the hostas brought from Japan to Holland by Siebold in 1829. It has been long-established in gardens in Europe and Britain and is used extensively in America as an edging for borders, along drives and round buildings. There are several botanical variants of this attractive, twisted-leafed hosta. It was described from material growing in Japanese gardens and has never been found in the wild and is, therefore considered as a 'species of convenience' or as an assemblage of garden clones. The sectional name 'Foliosae' alludes to the conspicuously variegated, sterile foliaceous bracts, which appear near the top of the scape, just below the raceme, especially if the plant has recently been disturbed or moved. It is a highly unstable variegated plant both in the long and the short term, and the crowns sometimes sport a green-leafed form, *H.u.* var. *erromena*, which is considerably bigger than the type. Some botanists consider this plant to be a species but it cannot be since it is sterile. *H.u.* var. *erromena* more usually arises from var. *univittata* than from *H. undulata*, thought not invariably. *H. erromena* is remarkable for the unvarying shape and size of its leaves, the leaves of its putative sports being notably inconstant in shape. Var. *erromena* has also arisen as a sport from *H.u.* var. *albo marginata* (*H.* 'Thomas Hogg'), but equally a white-margined sport has arisen from var. *erromena* that is slightly different from the familiar form.

The best form of *H. undulata* has creamy-white leaves with a narrow, irregular, exaggeratedly twisted and curled, dark olive-green margin. One peculiarity of this group is that the second crop of leaves differs from the first in having much less distinct variegation, often merely an unattractive streaking, and it is sometimes erroneously believed that the plant is losing its variegation, when in fact the new leaves the following spring will be as variegated as before. Siebold listed forms which he called *Funkia striata* and *Funkia argenteo-striata* and these were in all probability aberrant forms exhibiting this characteristic. They were subsequently dropped from the nursery catalogue. *H. undulata* and its forms and varieties are sterile even if artificially pollinated.

*

Medium-sized, clump-forming hosta, somewhat lacking in vigour in the United Kingdom. Sprouting shoots conspicuously dark reddish-purple, shiny. Petiole up to 5 in (13 cm), deeply grooved, striped green-white, purple-dotted at base, winged. Blades vary between ovate-oblong and lanceolate-elliptic, strongly undulate (more so than *H. crispula*), apex gradually acuminate, twisted through 180°, decurrent and often rolled at base; $4\frac{1}{2}$–6 in (11–15 cm) long × $1\frac{1}{2}$–$2\frac{1}{2}$ in (4–6 cm) wide; of thin substance but leathery texture; creamy-white centre, mid-to-dark olive-green irregular margin, underside shiny with 7–9 pairs of veins. Scape creamy white, finely lined green, purple-tinged, slightly grooved to the touch, up to 12 in (30 cm). Foliaceous bracts navicular, conspicuously variegated green-white, $3\frac{1}{2}$ in (7.5 cm) long, forming from centre of scape upwards, occurring more often if the plant is disturbed. Flower bracts white, margined green, withering after anthesis. Flowers pale violet, 30–40 per raceme. Anthers violet. Sterile.
Habitat: Gardens in Japan.

Illustr.: *Gentes Herbarum*, 2, 136, 1930; *AHS Bulletin*, No. 3, pp. 11–13; *Hosta: The Flowering Foliage Plant*, Fig. 13.

Var. *univittata* (Miq.) Hyl. Listed in Siebold's catalogue since 1863. Blade larger, more ovate, less undulate at margin but with a distinct twist. Central variegation more clearly defined, only $\frac{3}{4}$ in (2 cm) wide. After flowering, new leaves with less distinct variegation appear. Poise of leaves make a downward-rounded mound. Scape well overtops leaf-mound. A striking and useful garden plant.
Illustr.: *Hosta: The Flowering Foliage Plant* (Fig. 3).

Many intermediates exist between the extremes of *H.u.* var. *univittata* and *H. undulata*. The epithets *H.u.* 'Medio-picta' and *H.u.* 'Medio Variegata' are often used in the nursery trade but have no botanical standing.

H.u. var. *albo-marginata* (Maekawa 1936). Leaf blade ovate-elliptic, blade flat, margins slightly rippled, apex less acute than type; equal in size of *H.u.* var. *univittata*, with 9 or more pairs of veins; mid-olive-green upper surface with irregular white margins. Sterile foliaceous bracts, diminishing in size, arranged up the scape which well overtops the leaf-mound. Known in Britain as *H.* 'Thomas Hogg' (p. 29) and one of the most commonly-grown hostas in British gardens.

H.u. var. *erromena* (from the Greek, 'vigorous', or 'healthy') (Stearn) = *H. japonica* (Thunberg) var. *fortis* (Bailey). Green-leafed sport, considerably larger in size and more vigorous than the type. Blade broadly ovate, margin rippled, surface slightly leathery, dull, mid-green, 7–9 in (18–23 cm) × $5\frac{1}{2}$ in (14 cm) wide with 8–9 pairs of veins. Scape up to 44 in (120 cm), round, green, with 2 or 3 foliaceous bracts near raceme. Grown to best effect in a large planting.
Illustr.: *Hosta: The Flowering Foliage Plant*, Fig. 12, Pl. 34.

Many named hybrids from *H. undulata* exist but most are unstable.

4.

HOSTA CULTIVARS

There is a continual flow of new cultivars coming onto the market while, to a lesser extent, old cultivars and unstable or unsatisfactory ones tend to drop out of sight as they are superseded. There is no attempt here to list every hosta cultivar ever raised or registered (there are now some 1,000) but rather to concentrate on those which seem good garden plants. The majority of these are offered for sale by some of the larger British and United States nurseries and also by several small specialist hosta nurseries (see p. 200), and all have been grown for several years in different garden situations. Also included are some of the early cultivars which are of historic or botanical interest or are merely collectors' plants, and this is indicated in the description.

In the past five years, since the advent of tissue culture propagation as big business, there has been a veritable explosion of new hostas coming onto the market. Disappointments, however, are still frequent because many appear before they have been properly tried and tested and, indeed, often before even a single plant has been grown to maturity. It is too soon to judge many of these plants. If one raised a hosta from seed one would wait 3–5 years before making even the most tentative judgements on its merits and there is no reason why hostas arising as sports during the process of tissue culture, or new hybrids immediately sent to tissue culture laboratories, should be judged differently. It is how they perform in the garden that matters. Many test-tube cultivars have disappeared after only the briefest debut and others have proved unsatisfactory in one way or another.

Furthermore micropropagation seems to produce new problems – 'melting out' being one. This is a problem that occurs on some hostas having large areas of white variegation on the leaf blade; scorching appears and then a hole through the leaf. This can happen even if the plant is grown in shade. And it does not necessarily affect only hostas with thin leaves. The two hostas most commonly known to be affected in this way are *H.* 'Reversed' and *H.* 'Celebration' whose potential would otherwise be enormous.

As well as trying to breed for ever newer and different patterns of leaf variegation, hybridizers have made improvements over the years in the size and poise of the flower and the height of the scape in relation to the size of the leaf-mound; they are now producing gold-leafed cultivars with white flowers and fragrance in a much wider range of cultivars. Some hybridizers are concentrating on improved leaf shape, including such factors as ruffled margins (p. 120). Whilst one must applaud and encourage progress – and hostas have come a long way since *H. lancifolia* was first introduced to the West – the thought of flowers as gaudy as some hemerocallis flowers, growing out of a giant sized, multi-variegated leafed hosta, fills one with dread. It is hard to see how one could improve on cultivars such as *H.* 'Krossa Regal', *H.* 'Sum and Substance' and *H.* 'Green Fountain', with the simple flower counterbalancing the classic elegance and attractive colour of the leaf – but the raisers say they can.

It is increasingly the practice of American and now British nurserymen to express anything that is not a species as if it were a cultivar – that is, to give it a fancy name. The disadvantage of this procedure is that it tells nothing of the botanical standing of the plant, and therefore one can deduce nothing of its appearance or its cultural requirements from its name, nor indeed its breeding potential; all of which information is latent in the name when it is expressed as a variety, form or sub-species where applicable. The term cultivar properly applies to varieties which have arisen in cultivation and this is basically the rule followed here, though there are anomalies in hosta since some properly published species turn out to have been described from garden material. For those more familiar with fancy names cross-references have been included.

H. 'Allan P. McConnell' (Seaver)

Named by Mildred Seaver for a hosta-grower friend. Small, attractive, tight mound-forming clump. Petiole 5 in (13 cm) long, grooved, purple-dotted on the back. Blade ovate-cordate, matt, dark green with narrow, irregular white margin $2\frac{1}{2} \times 1$ in (6 × 2.5 cm). Scape 8 in (20 cm), dark purple, small fastigial bract with white margin, lavender flowers in midsummer. Good front-of-the-border plant.

H. 'Amanuma' (Maekawa)

Erroneously thought at one time to be a species because it arrived in America from Japan with a Latinized name. Now thought to be a seedling of either *H. nakaiana* or *H. capitata*. Named by Dr Maekawa in 1940 after his own house. Many *H. nakaiana* hybrids exist and most are good garden plants, usually remontant if the climate is warm enough. See *H.* 'Pastures New', p. 118.

H. 'Anne Arrett' (Arett-Minks)

Related, and similar to *H.* 'Kabitan'. Flattish-growing, slightly stoloniferous open clump.

Petiole deeply grooved, ruffled and white-margined. Blade decurrent, pale yellow, darkening slightly in autumn, ruffled white margin which develops with maturity. Not very vigorous. Needs shade.

H. 'Antioch' (Tompkins-Ruh-Hofer)

One of several similar selected forms of *H. fortunei* var. *aureamarginata* with cream to white-margined leaves. Very similar to *H.f.* 'Albo-aurea' nomen. illegit. (now *H.* 'Spinners', p. 122). Originally listed as *H.f.* 'Aurea-marmorata' (yellow-marbled) on account of the slight marbled section of the blade between the green centre and the creamy-white margin. First distributed by the old Wayside Nursery, Mentor, Ohio. The name was changed to *H.* 'Antioch' in the late 1970s, this cultivar having been discovered in a nursery field at Wayside's Antioch Farm. Robust medium to large hosta. The leaves bend over almost to the ground. Blade ovate, graduating to a point. Matt, mid-to-dark green at the centre, surrounded by green mottling or marbling, the margins irregularly splashed cream, turning white towards mid-summer, especially if grown in the sun. Prominent veins. Lavender flowers in midsummer. Should be grown in partial shade to achieve best leaf colour. *H.* 'Goldbrook' looks almost identical but smaller.

H. 'Aoki' (Syn. aoki hort., *Funkia aokii hort*),

George Schmidt wrote in the Fall 1986 *AHS Bulletin* 'Not of Japanese origin, but evolved in Holland as a member of the *H. fortunei* complex during Siebold's lifetime. Named by him in honour of the Japanese author Aoki Koniyo who produced the first Dutch/Japanese dictionary in 1740. First listed in commerce by Siebold's nursery in 1879 and described as having "elliptic leaves and very light blue flowers tinged with pink". It was then distributed to other European countries and to the USA'. Blade up to 12 × 7 in (30 × 20 cm), cordate-ovate, pale green with a network of darker veins giving a slightly variegated appearance early in the season, becoming an even green towards anthesis. Lustrous on upper surface, thinly pruinose below with 8–10 pairs of veins. Scape up to 40 in (100 cm). Bracts thin and withered, yellowish green. Raceme compact with flowers light violet to pale whitish mauve.

H. 'Aspen Gold' (Grapes)

Seedling or sport of *H. tokudama*. One of the first of the gold-leafed hostas raised which remain gold throughout the summer. Blade larger than *H. tokudama*, very rugose, of heavy substance, cupped. Near-white flowers barely overtop leaf-mound. Very slow-growing but one of the best. *H.* 'Midas Touch' (Aden) is similar, as is *H.* 'Bengee' (Palmer) although nearer chartreuse than gold. Very slow.

H. 'Apple Court Gold' (Grenfell/ Grounds)

Selected seedling from gold-leafed bed, presumed *H.* 'Semperaurea' × *H. tokudama* 'Flavo-circinalis'. Large hosta of heavy substance. Petiole 6in (15 cm), shallowly grooved. Blade ovate-cordate, slightly undulate, rugose, prominently veined, margins minutely saw-toothed with fine white pencil line, elongated tips, 7 × 5 in (18 × 13 cm). Light green at first, turning bright, brassy gold until frost. Short scape, shaded purple with several foliaceous bracts. Flowers light pinkish-mauve in mid-summer.

H. 'August Moon' (Summers)

Found in a garden in Westchester County, New York in 1964. Origin unknown, although a similar plant appeared in a garden in Connecticut at about the same time. Still one of the best gold-leafed cultivars. Medium to large, vigorous clump. Blade up to 9 × 3½ in (23 × 9 cm), slightly rugose, slightly pruinose, unfurls pale green and gradually becomes soft, pale gold. Will withstand full sunlight in Britain. Pale lavender flowers on tall scapes in July.

Repeat flowering when established. Increases rapidly. *H.* 'Lunar Eclipse', *H.* 'Mayan Moon' and *H.* 'September Sun' are all attractive sports from *H.* 'August Moon' but still very rare.

H. 'Beatrice' (Williams)

Seedling from *H. sieboldii*. Small to medium clump. Blade all-green but the whole plant is genetically unstable and has the unusual quality of producing a high proportion of variegated seedlings. It has been used extensively by hybridizers endeavouring to raise hostas with unusual variegated leaf patterns. Not outstanding as a garden plant.

H. 'Betsy King' (Williams)

Seedling from *H. sieboldii* with, it is presumed, *H. decorata* as the other parent. Flowers similar to *H. decorata*, light to mid-purple, the colour solid except for 6 white lines at the juncture of the perianth segments. Small to medium clump, leaf-mound 14 in (35 cm). Scape up to 20 in (50 cm). One of the best of Mrs Williams's purple-flowered cultivars. Others are *H.* 'Purple Profusion', *H.* 'Sentinels', *H.* 'Slim Polly' and *H.* 'Lavender Lady' (pinkish flowers). Of historical interest as they are among the first of the American-raised cultivars. Still commercially available in the United States.

H. 'Big Daddy' (Aden)

An *H. sieboldiana* form. A large, vigorous hosta, with glaucous-blue, round, puckered and cupped leaves of heavy substance. The flowers are white in the United States but pale mauve in British gardens. This applies to a number of hostas and is probably dependent upon the summer temperatures. Requires some shade. Is said to be resistent to slugs and snails.

H. 'Big Mama' (Aden)

An *H. sieboldiana* selection with even bluer leaves than *H.* 'Big Daddy'. Its leaves are less round and more pointed at the tips. It is also a rapid increaser.

H. 'Blue Angel' (Aden)

An *H. sieboldiana* form. A large, vigorous hosta with cordate glaucous-grey leaves with only a hint of blue. Dense raceme of pale lavender (white in the United States) flowers in mid-summer. Takes at least three years to look anything at all and from then on its gets better and better. A super hosta, but it should not be grown next to bluer-leafed hostas, as its leaves will look dull and faded. It should be contrasted with a shiny, gold or green-leafed hosta.

H. 'Blue Boy' (Stone)

Possible seedling of *H. nakaiana*. Small-to-medium graceful mound. Blade cordate, of thin substance, slightly glaucous, with a bluish-grey cast. Increases rapidly. One of the first of the smaller, blue-leafed hybrids. Not often seen in gardens now.

H. 'Blue Dimples' (Smith)

One of the second generation of hybrids of the Tardiana group, but one of the first to be named and distributed. It is considered to be one of the bluest. The leaves are cordate, tapering to a fine tip $4\frac{1}{2} \times 4$ in (11.5×10 cm) and at maturity the margins are undulate and a dimpling appears on the surface. Young plants, which in common with most hostas of the Tardiana group are often fairly indistinguishable from each other, have been confused with *H.* 'Blue Wedgwood' and offered, wrongly identified, in the trade in both Britain and the United States.

H. 'Blue Mammoth' (Aden)

An *H. sieboldiana* form. The leaves a lighter blue than most of the named forms of *H. sieboldiana*. Very large leaves with pronounced puckering. The petiole is short, widely grooved, and sharply angled. It has one or more long, narrow foliaceous bract on its thick scape, suffused purple. Lavender flowers in mid-summer. An excellent back-of-the-border plant.

H. 'Blue Moon' (Smith/Aden)

(Originally named 'Halo'.) The smallest of the Tardiana group to be named. Blade cordate/orbicular, of good substance, margin flat, stiff, smooth glaucous blue, $2\frac{1}{2} \times 1\frac{1}{2}$ in (6×4 cm). Glaucous upright scape, up to 5 in (12.5 cm), barely over-topping leaf-mound, bearing dense raceme of hyacinth-blue flowers in mid-to-late summer. Very slow-growing. Suitable for a shaded position on the rock garden or peat bed. Needs dappled shade.

H. 'Blue Piecrust' (Summers)

H. sieboldiana seedling. Blade ovate-cordate, deeply cupped, less rugose than some, the margins ruffled like a piecrust, deep glaucous blue. Interesting as one of the first hybrids of this type, though others now have more pronounced pie-crust edges and similar forms exist with green leaves and with leaves of a greyish cast. A curiosity rather than a hosta of great distinction. S. also *H.* 'Ruffles' (p. 120).

H. 'Blue Seer' (Aden)

H. sieboldiana form. A large, but not one of the biggest blue-leafed hostas. Robust. Blade deeply puckered, of heavy substance, ovate, tapering to a point, glaucous turquoise blue-green. White or near-white flowers in mid-summer. Excellent specimen plant.

H. 'Blue Umbrellas' (Aden)

H. sieboldiana form. A very large hosta with broad, greenish-blue glaucous leaves with widely spaced veins. Of thick substance and highly resistant to slug and snail damage.

H. 'Blue Wedgwood' (Smith/Summers)

There is some confusion over which of the Tardiana group *H.* 'Blue Wedgwood' actually is, and some nurseries may be offering the wrong hosta under this name, a mistake that is easily made since young plants of hostas of the Tardiana group differ substantially in leaf size and shape from mature plants. However, the description given below accords with Eric Smith's own notes. There is also confusion as to how this hosta came by its name. The supposition was that Eric Smith thought the leaves similar in colour to blue Wedgwood pottery, but Alex Summers says that when he and Smith were choosing names for the original twelve Tardianas they thought this one had leaves like a wedge of wood and being blue they called it 'Blue Wedgewood' with a middle 'e'. It has been registered without the middle 'e'. A medium-sized open clump. Blade triangular with rippled margin, $6\frac{1}{2} \times 3$ in (26×8 cm), smooth, glaucous blue-green. Glaucous scape just above leaf-mound bearing raceme of pale lavender flowers in midsummer.

H. 'Bobbin' (O'Hara)

One of the smallest variegated hostas. An *H. venusta* seedling. Not yet commercially available.

Fig. 12 *H. undulata* var. *erromena*
Description on p. 100

H. 'Bold Ribbons' (Aden)

Related to *H.* 'Neat Splash' and *H.* 'Yellow Splash' and, like them, stoloniferous. Medium-sized clump of arching leaves attractively margined in creamy white. Blade ovate-lanceolate, matt mid-to-dark green, tapering to a slightly rounded tip; irregular wide margin. A striking plant, but could be confused in its juvenile state with *H.* 'Wide Brim'. Useful for flower arranging.

H. 'Bressingham Blue' (Bloom)

Probably a hybrid between *H. sieboldiana* 'Elegans' and *H. tokudama* exhibiting the best qualities of both. Wide, cup-shaped, rugose leaves of heavy substance, glaucous deep blue-green. Pearly-white flowers in midsummer. One of the best.

H. 'Bright Glow' (Aden)

Creamy gold-leafed Tardiana seedling. White flowers in July. Of paler colour and thicker texture than the golden Tardiana given to the author by the raiser. The leaf shape is similar.

H. 'Buckshaw Blue' (Smith) (AM)

Award-winning seedling of what is presumed to be *H. tokudama* but correspondence from Eric Smith to Alex Summers mentions that it might be a hybrid between *H. sieboldiana* and *H. tokudama*. Found in the 1960s at Hilliers Nursery, Winchester and later taken to The Plantsmen Nursery by Eric Smith and named after its premises, Buckshaw Gardens. Noticed a few years later by the late Ken Aslett who had a piece planted at Wisley Garden. It was subsequently given an Award of Merit. Similar to *H. tokudama* but selected on account of its deeper blue, larger and more dished leaves. Greyish-lavender flowers in July do not open wide. Scape barely overtops leaf-mound. Not a vigorous or easy plant to establish and it seems to require hotter temperatures than we have in Britain to achieve its potential. In 1987 it received the Nancy Minks Award for the best small to medium-sized hosta on the American Hosta Society convention tour that year. An enormous clump was growing in Alex Summers's garden in southern Delaware, proving the point that it grows best in hotter climes.

H. 'Bunchoku'

Dwarf form of *H. sieboldii* with prominent white margin, the blade merging with the petiole (decurrent). Popular as a pot plant in Japan. Sometimes erroneously known as 'Chisima giboshi' or 'Shirofukurin Chisima giboshi'.

H. 'Butter Rim' (Summers)

Seedling of *H. decorata* and *H. sieboldii*. Blade up to $4 \times 1\frac{1}{2}$ in (10 × 4 cm), matt, dark green, bordered by creamy-yellow margin. Good white flower in late summer. One of the very few hostas which is usually not a 'good doer' in Britain, except in hot, sunny summers. It is a vigorous plant in the USA.

H. 'Carol' (Williams)

Named by Mrs Williams for her granddaughter. Exact origin uncertain, but thought to be either a seedling with some *H. fortunei* in its parentage, or a bud sport. One of the many white-margined cultivars of *H. fortunei* extraction. Robust, medium-sized clump. Blade ovate, 7×3 in (18 × 8 cm), glossy, mid-to-dark green with irregular and unstable white margin. Not a rapid increaser and needs good cultivation to grow well.

H. 'Chartreuse Wiggles' (Aden)

Small hosta with its leaves lying nearly flat to the ground. Leaves narrow, very undulate, pale chartreuse yellow. Purple flowers in late summer. Not vigorous in British gardens.

H. 'Chelsea Babe' (Smith/Grounds)

A curious little hosta that looks exactly like a very dwarf version of the Chelsea hosta, *H. fortunei* 'Albo-picta' but which is almost certainly derived from some other species since details of the flowers and petioles are quite wrong for an *H. fortunei*. Distributed by Eric Smith who may have received it from America. It may be *H.* 'Maya', (Summers) which is an American cultivar, so weak-growing that it seems to have disappeared from cultivation. Although *H.* 'Chelsea Babe' is now grown by several collectors in Britain it may not prove to be a worthwhile garden plant.

H. 'Chinese Sunrise' (Ruh)

Imported into America from Japan unnamed. Thought to be a variegated form of *H. lancifolia*. Small to medium stoloniferous hosta. Blade 5 × 1 in (13 × 2 cm), margin flat, tapering to a fine point, of thin substance, glossy, chartreuse to dull-gold centre with narrow, dark green margin. Gold fades to dull green during the summer. Dark lavender flowers in August and September.

H. 'Claudia' (Grenfell)

Named for author's daughter. Cross between *H. sieboldii* and *H. clausa* forma *normalis* which accounts for its beautiful deep purple flowers. Spatulate leaves, up to 1 in (2.5 cm) wide, 6 in (15 cm) long. Shiny, bright mid-green. Flowers late summer.

H. 'County Park' (Hutchins)

Found by Roger Grounds in a batch of *H. elata* seedlings. Medium-sized clump, that looks for all the world like a cos lettuce. Leaves densely bunched at the centre, outer leaves lying flat. Petioles 1½ in (4 cm) long, grooved. Blade almost orbicular, rugose, slightly cupped, 5 × 3½ in (13 × 9 cm), matt light to mid-green. White bell-shaped flowers on short scapes. Of curiosity value on account of the very distinct shape of the clump, but liable to suffer from crown rot.

H. 'Devon Blue' (Smith/Bowden)

Named and distributed by Ann & Roger Bowden who acquired it when they took over the late John Goater's nursery. It had been sold to him by Jim Archibald at The Plantsmen as '× tardiana F1 hybrid'. A medium-sized, vigorous hosta, growing equally well in a pot or open ground, in sun or shade. Blade ovate-cordate, tapering to a point, veins widely spaced 7 × 4 in (18 × 10 cm), glaucous, greenish-blue-grey. After 5 years clump is 18 in (45 cm) tall × 30 in (90 cm) wide. Glaucous scape up to 30 in (90 cm). Flowers light violet in mid-summer.

H. 'Dorset Blue' (Smith)

One of the most rugose-leafed of the Tardiana group, reminiscent in leaf shape and texture of *H. tokudama*. Small, very slow-growing hosta. Blade cordate-orbicular, rugose, of heavy substance, glaucous, deep turquoise-blue, 5 × 3½ in (13 × 9 cm). Stout, glaucous scape, barely over-topping leaf-mound, bearing raceme of pale greyish-lavender flowers in midsummer. One of the best of the Tardiana group.

H. 'Elizabeth Campbell'

(Raiser not known), syn. *H.* 'Panache Jaune' (meaning yellow variegated). Thought to be an improved form or sport of *H. fortunei* var. 'Albo-picta'. In early summer it is one of the most attractive variegated-leafed cultivars and superior to the similar *H.* 'Phyllis Campbell' with which it is often understandably confused. A flower arrangers's plant. Open clump. Petiole 8 in (20 cm) long, shallowly grooved, winged, variegation at centre of leaf extends down centre of petiole; petiole pinched at junction with blade. Blade larger and more cordate than *H.f.* 'Albo picta', undulate, slightly twisted tapering tip, margins slightly ruffled; creamy-yellow centre 1–3 in (2.5–8 cm) wide with irregular mid-to-dark green margin. The central variegation becomes wider and more pronounced as the plant matures. In common with other sports arising from hostas of *H. fortunei* origin the variegation fades after midsummer, but to a lesser degree in this hosta than most. Scape well overtops leaf-mound and is purple towards the top: foliaceous bracts wither on anthesis, large dense raceme of mauve flowers.

H. 'Ellerbroek' (Ellerbroek)

A hosta originally selected by Dr Hylander (who named it *H. fortunei* var. *obscuro marginata*, syn. *H.f.* var. *aureo marginata*,) and given to Dr Hensen who, in turn, sent a piece to Paul Ellerbroek of Des Moines, Iowa, with whom he was corresponding and exchanging plants, probably in the mid-60s. Probably just a good selected form of *H.f.* 'Obscuro marginata'.

H. 'Emerald Isle' (Smith)

Similar to *H.* 'Louisa' (Pl. 31) but smaller and even less vigorous. A white-margined form of *H. sieboldii* with white flowers.

H. 'Emerald Skies'

(Raiser not known), small, front-of-the-border hosta. Shiny, thick-textured, very dark green leaves. Flowers mauvish. Very slow-growing. A sport from *H.* 'Blue Skies' which arose through tissue culture.

H. 'Eric Smith' (Smith/Archibald)

This is the hosta thought by Eric Smith himself to be the best of the Tardiana group and he is reputed to have told his partner at The Plantsmen, Jim Archibald, that if ever anyone wanted to name a hosta after him, this was the one he would like them to name. Originally known as TF 2 × 31. Small, dense clump. Blade cordate, margin shallowly undulate, matt, glaucous blue-green (darker in colour than most of the Tardiana group). Scape overtopping leaf-mound by about 3 in ($7\frac{1}{2}$ cm), bearing deep lavender flowers in July. Leaves parch in sun.

H. 'Feather Boa' (O'Harra)

Small, tight mound of pale yellow lanceolate leaves. Attractive on account of its exceptionally tall scapes, 20 in (50 cm) which tower over the leaf mound, bearing racemes of off-white flowers. Looks magnificent as an edging plant along a path.

H. 'Flamboyant' (Aden)

Purported to have been raised by x-ray treatment just before the advent of tissue culture. One of the first of those multi-variegated cultivars which change their character once the juvenile phase is over, usually after about 3 years. The adult form lacks the central streaked variegation of the juvenile phase, and has instead a green leaf with a fairly wide $\frac{1}{4}$ in (1 cm) cream margin. The juvenile form is known as *H.* 'Flamboyant' but the adult form is *H.* 'Shade Fanfare' (p. 121). The rule, laid down by the International Commission for the Nomenclature of Cultivated Plants, which follows the rules of the ICBN, is that hostas which quite plainly appear to be different must bear different names. *H.* 'Flamboyant' has become the case hosta illustrating the rule. Medium-sized open clump. Blade cordate-ovate, margins shallowly undulate, tips recurved, 6 × 4 in (15 × 10 cm), glossy, mid-green, splashed and streaked cream, yellow, chartreuse, gold. Pale lavender flowers in mid-summer. Very floriferous. Half to three-quarters shade required to produce best colour effects. Margin tends to bleach in full sun.

H. fluctuans 'Variegated'

See p. 72.

H. fortunei 'Albo-picta'

See *H. fortunei* var. *albo-picta* (p. 79).

H. fortunei 'Albo-marginata'

See *H. fortunei* var. *obscura* (p. 79).

H. fortunei 'Aurea'

See *H. fortunei* var. *albo-picta* forma *aurea* (p. 79).

H. fortunei 'Aureo-marginata'

See *H. fortunei* var. *aureo-marginata* (p. 79).

H. fortunei 'Hyacinthina'

See *H. fortunei* var,. *hyacinthina* (p. 78).

H. fortunei 'Rugosa'

See *H. fortunei* var. *rugosa* (p. 78).

H. fortunei 'Stenantha'

See *H. fortunei* var. *stenantha* (p. 78).

H. 'Francee' (Klopping)

Selected form of *H. fortunei*. Neat, clump-forming, medium to large hosta, suitable for the middle of the border. Leaf-mound up to 10 in (25 cm). Blade ovate-orbicular, slightly cupped, slightly rugose, 7 × 3 in (18 × 8 cm), mid-to-deep green, bordered by regular, narrow clear-white margin. Scape up to 24 in (60 cm), bearing racemes of pale lavender flowers in late summer. Grows in sun or partial shade. Increases rapidly.

H. 'Frances Williams' (Williams)

The most popular hosta ever raised. For historical details see p. 110. Bud sport of *H. sieboldiana* 'Elegans'. Similar to *H.s.* 'Elegans' in leaf but slightly smaller. Blade cordate, rugose, of heavy substance, glaucous deep blue-green, bordered with wide irregular margin of creamy beige. Foliaceous bracts up to 3 in (8 cm) long, near raceme, still turgid at anthesis. Dense raceme on stubby scape of near-white flowers which just overtop leaf-mound, in midsummer. Takes at least four years to establish. Best grown in dappled shade. Plants of *H. tokudama* 'Flavo-Circinalis' are often mistaken for young plants of *H.* 'Frances Williams'. The leaves of *H.* 'Frances Williams' are much rounder and less pointed towards the tip. Like the type, *H.* 'Frances Williams' is very variable depending on cultural conditions. Many forms have been thought distinctive enough to name, (*H.* 'Golden Circles' and *H.* 'Eldorado' (Archibald)), but it is now generally accepted that all these forms are merely variants.

H. 'Friesing' (Mussel)

Raised at Weihenstephan Horticultural and Brewing Faculty's Trial Garden, Freising, Munich. A slow-growing selected form of *H.*
fortunei. A medium-sized hosta of stiff habit. Blade lanceolate-ovate, distinctly undulate margin, tapering to a point, 6 × 2½ in (15 × 5 cm), leathery texture, glaucous mid-green with slight blue cast. Scapes short, more or less upright. Attractive white bell-shaped flowers in midsummer. A white-margined form has occurred in Britain.

H. 'Fringe Benefit' (Aden)

A good background and landscaping hosta with characteristics of *H. sieboldiana* and the *H. fortunei* complex. Medium to large clump. Blade cordate-ovate, slightly rugose, shallowly undulate margin, dark green with a wide, irregular creamy-white margin, 9 × 9 in (22 × 17 cm). Pale mauve flowers in early summer. Will withstand full sun. A rapid increaser. Slug resistant.

H. 'George Smith' (George Smith)

Bud-sport from *H. sieboldiana* 'Elegans', raised by well-known flower arranger, George Smith. A large, slow-growing hosta with striking foliage. Slightly smaller than the type, much less vigorous but well worth a place in a collector's garden. Blade cordate, rugose, of heavy substance, pruinose, creamy-ivory with irregular blue-green margin, up to 12 × 6 in (30 × 25 cm) on mature plant. Off-white flowers in midsummer on short, thick scapes. This is the first hosta raised with the reverse variegation to *H.* 'Frances Williams', but many, slightly different forms are occurring, some through tissue culture, e.g. *H.* 'Northern Sunray' (Walters). *H.* 'Borwick Beauty', 'Great Expectations', 'Northern Sunray' and 'Color Glory' are all similar and it may be that under the rule of nomenclature that requires cultivars that look the same to bear the same name, these should all be called 'George Smith'.

H. 'Ginko Craig' (Craig/Summers)

Named by Jack Craig for his Japanese wife, Ginko. Small, front-of-the-border or edging plant. Mound exceptionally low-growing. Short flattish petiole, edged white. Blade decurrent, lanceolate, tapering suddenly, matt, olive-to-dark green, inner border marbled greyish green; narrow, irregular white margin which widens as the plant matures, $6 \times 1\frac{1}{2}$ in (15×4 cm). Scape 10 in (25.5 cm), bearing raceme of deep mauve bell-shaped flowers in mid to late summer. A ungainly hosta in its youth. A rapid increaser, in some gardens. Needs some shade.

H. 'Gloriosa' (Krossa/Summers)

Selected form of *H. fortunei* complex. Small to medium-sized hosta. Blade spatulate, slightly concave, furrowed, 6×2 in (15×5 cm), matt, mid-to-dark green with pencil-thin clear white margin which sometimes disappears altogether (more distinct and slightly wider on mature plants). Slightly stoloniferous. Lavender flowers in midsummer. Not widely available in Britain.

H. 'Gold Bullion' (AHS Nomenclature Committee) (p. 77)

The cultivar or clonal name registered by The American Hosta Society's Nomenclature Committee to cover all gold-leafed sports arising from *H. tokudama* 'Flavo-circinalis'. Some growers suspect that *H.t.* forma 'Flavo-circinalis' might not be as closely related to *H.t.* 'Aureo-nebulosa' as Maekawa (1940) suggests, as its leaves are longer and more pointed than the type. This has also been borne out in the shape of the gold-leafed progeny arising from tissue culture. Different chloroplast structures have also been observed under an electron microscope which adds weight to the theory. However it has been observed that from a plant of *H.t.* 'Aureo-nebulosa' growing in a British garden a bud sport appearing to actually resemble *H.t.* 'Flavo-circinalis' has arisen.

H. 'Gold Drop' (Anderson)

Seedling of *H. venusta*. A dwarf hosta suitable for the rock garden or peat bed. Blade cordate-ovate, dull gold, heavy-textured leaves which

need some sun to attain a reasonably good colour. Scapes 5 in (12.5 cm) bearing raceme of attractive mauve flowers in midsummer. *H.* 'Crown Jewel' ('Walters Gardens') is a cream-edged, pale green-leafed sport.

H. 'Gold Edger' (Aden)

Small edging or front-of-the-border hosta which increases rapidly to produce dense ground cover. Can easily be mistaken for *H.* 'Birchwood Parky's Gold' which has thinner textured leaves and longer petioles. Blade ovate-cordate, of heavy substance, smooth surface, slightly pruinose, up to 4×3 in ($10 \times 7\frac{1}{2}$ cm). Reasonably sun-tolerant and slug and snail resistant. Masses of lavender flowers in midsummer. A very good hosta.

H. 'Gold Haze' and H. 'Gold Leaf' (Smith)

Improved forms of *H. fortunei* 'Aurea'. Two of many seedlings raised by Eric Smith in an attempt to produce gold-leafed hybrids which remain gold until the frost (p. 55). See also *H.* 'Oriana' (p. 118). Most of them not entirely satisfactory as colour fades although not as early in the summer as in *H.f.* 'Aurea'. *H.* 'Gold Leaf' has a faint pruinose bloom overlying the gold. Both have very pale lavender flowers in mid-to-late summer. They look good when growing in association with magenta and other strong shades of the evergreen azaleas, particularly *Rh. amoenum obtusum*.

H. 'Gold Regal' (Aden)

Large, vigorous clump with stiffly erect leaves. Owes more in stature to *H. rectifolia* than to *H.* 'Krossa Regal' whose gold-leafed counterpart it was once thought to be. Stiff, upright glaucous petioles, deeply grooved at the base and widening out towards the leaf junction, 12 in (30 cm). Blade ovate, slightly dished, 10×4 in (25×10 cm), margin flat, heavy substance, chartreuse/dull gold with a slight glaucous bloom. Tall, thick scapes topped with dense raceme of bell-shaped, ashen-mauve flowers in midsummer.

Foliaceous bracts wither at anthesis. One of the best hostas for flowers. Needs several years to develop before showing its full potential. Lacks the poise and grace of H. 'Krossa Regal' but still a fine garden plant.

H. 'Gold Standard' (Banyai) A.M.

One of the most popular of the newer cultivars to come onto the market and deservedly so. A sport from H. "glauca" (presumed to be H. fortunei 'Hyacinthina' – which is known to produce variegated sports) which, in turn, has produced its own sport, H. 'Moonlight' (p. 117). Typical H. fortunei in shape and size. Robust, fast-increasing medium to large open clump, less stiff than H. 'Moonlight'. Blade unfurls chartreuse-green with irregular deep-green margin. If grown in sufficient sun the blade will gradually turn bright gold, giving a vivid contrast to the dark green margin. If grown in too much sun, however, the leaf centres will turn papery and white and become far less attractive. A very striking garden plant if sited correctly and given the right cultural conditions. A flower arranger's plant. Very popular in Britain and the USA.

H. 'Golden Isle' (Smith)

Slightly larger and more vigorous than H. 'Emerald Isle' (p. 109). A selected form of H. sieboldii. This, and H. 'Emerald Isle' may well have counterparts under different names in Japan. Blade chartreuse yellow in early summer, with narrow mid-to-dark-green margin. Tall, slender scape bears raceme of white trumpet-shaped flowers in late summer. Needs a certain amount of pampering to achieve its full potential, but a delightful plant for the rock garden, peat bed or front of the shaded shrub border, grown in association with evergreen azaleas.

H. 'Golden Medallion' (AHS Nomenclature Committee)

The cultivar or clonal name selected by The American Hosta Society's Nomenclature Com-

mittee to cover all gold-leafed sports arising from H. tokudama, and H.t. 'Aureo-nebulosa' and 'Flavo-planata' either from naturally occurring bud sports or by callusing during tissue culture.

H. 'Golden Prayers' (Aden)

Seedling of H. 'Golden Waffles' (itself one of the first golden sports of H. tokudama now submerged in 'Golden Medallion'). Small to medium-sized hosta with leaves poised in an upright, incurved attitude, reminiscent of hands in prayer. Blade ovate-cordate, rugose, heavy substance, cupped, $5 \times 3\frac{1}{2}$ in (13×9 cm), pale to mid-yellowish-gold. Short, stout scapes, bearing off-white flowers in midsummer. One of the best of the smaller gold-leafed hostas.

H. 'Golden Sculpture' (Anderson)

A very large hosta. Huge, cordate leaves, smooth surface, of good substance, bright, pale gold, veins widely spaced. One of the new breed of highly sun-tolerant, pest resistant gold-leafed hostas. (H. 'Jim Cooper' [H. 'Sum and Substance' × unknown seedling] is similar). White trumpet-shaped flowers on a 36 in (115 cm) scape, makes a dense mound 24 in (70 cm) high and keeps its gold colour in sun or shade. Well worth looking out for as nursery stock becomes more available.

H. 'Golden Sunburst' (AHS Nomenclature Committee)

The cultivar or clonal name selected by the American Hosta Society's Nomenclature Committee to cover the gold-leafed sports arising from H. 'Frances Williams' either from naturally occurring bud-sports or by callusing during tissue culture.

H. 'Golden Tiara' (Savory) (AHS Nancy Minks Award)

A very pretty hosta, and one of the best and most successful of the newer small hybrids. According to Robert Savory, its raiser, in the

AHS Journal 1985 he 'treated 750 *H. nakaiana* seedlings with a mixture of hormones and vitamins in order to "break" more dormant eyes and to possibly stimulate mutations ... when older, dormant eyes are forced into growth, their shoots frequently provide a slightly higher rate of mutation than ordinary crown cuttings'. *H.* 'Golden Tiara' was one of these artificially induced sports. Small, compact, clump-forming, increases well. Blade cordate, pale to mid-green, margined with $\frac{1}{4}$ in ($\frac{1}{2}$ cm) band of chartreuse-yellow. Tall, slender scapes bearing raceme of lavender-purple flowers in mid to late summer. *H.* 'Golden Scepter', is a naturally-occurring gold-leafed bud-sport. The name *H.* 'Golden Scepter' is also used for the tissue-culture gold-leaf sports of *H.* 'Golden Tiara'. Through tissue culture, other leaf variegations have arisen and the group is called the 'Tiara' Series (*AHS Bulletin* Fall 1986 and *BHHS Bulletin*, Vol. 2, No. 2). 'Platinum Tiara' (gold with a white edge) (Walters Gardens), 'Royal Tiara' (T & Z Nursery), 'Silver Tiara' (Walters Gardens), 'Emerald Tiara' (Walters Gardens) and 'Diamond Tiara' (T & Z Nursery). None of these tissue-culture sports has yet been grown in gardens for long enough to evaluate their true worth.

H. 'Granary Gold' (Smith)

In the late 1970s Eric Smith gave George Smith, the well-known flower arranger, a piece of a gold-leafed hosta which was growing by the door of his cottage at Hadspen, Somerset. The cottage was called Granary Cottage, so Eric Smith called this hosta 'Granary Gold'. George Smith uses it in many of his arrangements.

It is a sport, vastly superior to its parent, *H. fortunei* 'Aurea'. It is one of the GL series described on p. 55. It has larger, more cordate leaves which retain the pale gold colour for most of the summer, gradually fading to creamy-green. It should be regarded as the true *H.* 'Granary Gold'. Other gold-leafed hostas are in circulation in British gardens under this name and the British Hosta and Hemerocallis Society

will have to address the question of their identity.

H. 'Green Acres' (Geissler)

A popular hosta when it was introduced in the 1960s although it will probably be superseded by the even larger *H.* 'Mikado' (Aden) and *H.* 'King Michael' (Summers) (Pl. 22). A large hosta forming an open, spreading mound. Suitable for the back of the border or as a specimen. Thought to be of *H. montana* origin. Petiole 12 in (30 cm) long. Blade 15 × 7 in (38 × 15 cm), ovate-lanceolate, matt, smooth surface, light sage-green, veins deeply impressed. Scape up to 20 in (50 cm) fastigial and flower bracts wither at anthesis. Lavender flowers in late midsummer.

H. 'Green Fountain' (Aden)

A selected seedling from *H. kikutii*. Larger than the type. Leaves cascade out from the base of the mound and arch over to the ground. Petiole spotted deep purple at the base. Blade 10 × $3\frac{1}{2}$ in (25 × 9 cm), tapering to the tip, margin more undulate than the type, glossy, mid-green, with prominent veins. Scapes bent almost horizontal, with racemes of very pale lavender flowers in late summer.

Fig. 13 *H. undulata*
Description on p. 100

H. 'Green Gold' (Mack/Savory)

A selected form of *H. fortunei* 'Aureo-marginata'. Medium-sized clump. Blade ovate-cordate, 7 × 4 in (18 × 10 cm), leathery texture, lustrous dark green, narrow irregular cream to light-yellow margin, less prominent towards the tip. Not as striking as *H.* 'Antioch', 'Ellerbroek' or 'Spinners'.

H. 'Green Wedge' (Aden)

An improved form of *H. nigrescens* var. *elatior*. A vigorous, large, clump-forming hosta. Blade wedge-shaped, 12 × 10 in (30 × 35 cm), glossy, light greenish-gold. Tall, slanting scape carrying raceme of pale lavender flowers in mid-summer. A giant hosta suitable for background or specimen planting.

H. 'Ground Master' (Aden)

Medium-sized, slightly stoloniferous, ground-hugging hosta. Blade lanceolate-ovate $9\frac{1}{2}$ × $3\frac{1}{2}$ in (24 × 9 cm), shallowly undulate margin, matt, dark green with an irregular wide, white margin. Masses of lavender flowers in August. Suitable for planting on banks and for ground cover. (Pl. 23).

H. 'Hadspen Blue' (Smith)

Small to medium-sized open clump. One of the bluest. A good contrast with *H.* 'Nancy Lindsay' (Grenfell) (Pl. 24). Petiole $5\frac{1}{2}$ in (14 cm). Blade cordate, smooth, closely veined, margin flat, $5\frac{1}{2}$ × 3 in (14 × 8 cm), glaucous deep blue. Glaucous scape, just overtopping leaf-mound, bearing raceme of mid-lavender flowers in July. The emerging shoots have a distinctly red tinge.

H. 'Hadspen Heron' (Smith)

Small, low-growing clump suitable for the rock garden or peat bed. Blade lanceolate, undulate, margin shallowly rippled, surface smooth, thick texture, glaucous mid-blue-green, 5 × 1 in ($13 × 2\frac{1}{2}$ cm). Glaucous scape barely overtopping leaf-mound, bearing dense raceme of lavender flowers in midsummer. One of the first to emerge, liable to frost damage.

H. 'Hadspen Samphire' (Smith/Eason)

Seedling of *H. sieboldiana* × *H.* 'Kabitan'. Slow-growing open mound of ovate-lanceolate, thin-textured, harsh, mid-gold leaves, fading to chartreuse, veins widely spaced. Must be grown in shade. Not very vigorous and slow to establish and increase. Named for the gold colouring reminiscent of samphire, a fleshy European seaboard plant of the parsley family. Many other seedlings of this cross were raised, about half of them with green leaves. *H.* 'Goldsmith' was another good hosta but it has disappeared from cultivation, (p. 55). *H.* 'Hadspen Honey' is also this cross but plants of *H.* 'Golden Sunburst' did, at one time, get passed around mistakenly with the name *H.* 'Hadspen Honey' and there is still some confusion concerning Eric Smith's intended names for this cross and his GL series.

H. 'Haku-Chu-Han' (Davidson)

Small, weak-growing form of *H. sieboldii*. The name translates from the Japanese as 'Half of leaf white in the centre'. Now considered likely to be synonymous with Maekawa's *H. lancifolia thunbergiana* f. *mediopicta*. Petiole 5 in (13 cm) long, white with narrow dark green margins. Blade spatulate, rather shapeless, somewhat decurrent, white in spring, becoming slightly off-white towards summer, margin olive-to-dark green. Of interest historically and of interest to collectors but not a particulary outstanding garden plant. It resembles hostas of the *H. undulata* species. Needs full shade.

H. 'Halcyon' (Smith) AM. Award of Merit Hosta, AHS 1987.

The best-known and most popular of the Tardiana group, (p. 123). Many seed-raised plants offered for sale under this name are merely poor imitations. It should be propagated vegetatively if the true plant with its very distinctive shape and colour is to be maintained. In common with many hostas, and in particular the Tardiana group, in its juvenile stage the leaf blade is almost lanceolate but as the plant matures it gradually becomes truly cordate. Margin flat, smooth, of medium substance, veins close together, $6 \times 3\frac{1}{2}$ in (15×9 cm). Glaucous scape just overtops leaf-mound, which bears exceptionally dense raceme of deep lavender flowers in mid-to-late summer. Leaf colour remains at its bluest if grown in some shade.

H. helonioides 'Albo-picta'

See *H. helonioides* f. *albopicta*, (p. 93).

H. 'Honeybells' (Cumming)

A hybrid originally thought to be between *H. plantaginea* and *H. lancifolia*, but this was a misidentification and the second parent was, in all probability, *H. sieboldii*. Generally nearer to the *H. plantaginea* parent. One of the first American hybrids to be a commerical success. Introduced to Britain in the 1960s. Blade virtually identical in shape, colour and texture to *H. plantaginea* 'Grandiflora'. The flower is white, streaked with violet, sweetly scented, though not so strongly as *H. plantaginea*. A rapid increaser, so excellent for covering large areas of ground. Will tolerate more direct sunlight than most.

H. 'Hydon Sunset' (George)

A hybrid between *H. gracillima* and H. 'Wogon Gold' raised by Arthur George of Hydon Nursery, Godalming, Surrey. A small cordate/ovate leafed hosta which emerges in the spring a bright, deep gold and gradually fades to a dull green. Deep purple flowers on tall petioles in mid-to-late summer. Often, understandably confused with *H*. 'Sunset' a smaller, but similar gold-leafed hosta which was misidentified. *H*. 'Sunset' is a slightly paler gold which keeps its gold colour until the frosts but is of thinner substance. Both these hostas require full shade.

H. 'Invincible' (Aden)

One of the very best of the new cultivars. Small to medium-sized open clump. Leaves lanceolate,

tapering to a fine tip, glossy and of thick substance, mid green, margins markedly undulate, up to 10 in (25.5 cm) long. Scape 8 in (20.5 cm) bearing dense raceme of pale lavender, scented flowers in late summer. Requires some shade; slug and snail resistant.

H. 'Janet' (Shugart)

Medium-sized, somewhat weak-growing hosta. Similar in leaf shape to the *H. fortunei* complex, blade 9×3 in (23×8 cm), matt, of thin substance; leaf colour changes from chartreuse to ivory to white, with narrow, dark green margin. By midsummer the blade will be parchment white. Needs careful placing between sun and shade to achieve the best colour. A striking hosta when well suited. Increases only very slowly and sometimes hardly at all.

H. 'Julie Morss' (Morss)

Introduced by British hosta grower and former nurserywoman, Mrs Julie Morss, one of the first British hosta enthusiasts. A sport from a grey-leafed seedling of *H*. 'Frances Williams'. Slow-growing, medium-sized clump. Blade ovate-cordate, convex, with tip turned under, $6 \times 4\frac{1}{2}$ in (15×11 cm), rugose, pruinose, light gold with wide, irregular glaucous-blue margin. Glaucous scape 18 in (45 cm), bearing lilac-pink flowers in mid-to-late summer. Leaves look best early in the season, fading slightly after midsummer. It needs dappled shade to retain the best colour. One of the best hybrids raised in Britain.

H. 'Kabitan'

See *H. sieboldii* forma *kabitan*, (p. 97).

H. 'Kasseler Gold (Klose)

Selected seedling from *H*. 'Semperaurea', raised by Heinz Klose the West German nurseryman and hosta collector. Named after the town of Kassel where his nursery is based. A large hosta with shiny, thinner, less rugose leaves than the parent. Needs protection from direct sunlight.

H. 'Krossa Regal' (Krossa Summers)

One of the classic hostas. Imported into the United States from Japan as a seedling by the late Gus Krossa, one of the first American hosta collectors, and named for him by Alex Summers. A tall, graceful vase-shaped clump, holding the leaf blades at a most distinct angle atop tall petioles. Petiole 10 in (25 cm), glaucous. Blade ovate-lanceolate, up to 12 in (29 cm), margin shallowly undulate, heavily pruinose, of good substance, the veins widely spaced, greyish-blue. Undulating scape up to 5 ft (150 cm), showing affinity with *H. nigrescens*, bearing raceme of orchid-pink, bell-shaped flowers in late summer. Superb, both in leaf and flower. A plant for flower arrangers.

New tissue culture sports of this hybrid have been named. 'Regal Spendour' (Walters Gardens) with cream margin, 'Porcelain Vase' (T & Z Nursery) with a central cream variegation are offered in the trade in the United States.

H. 'Lemon Lime' (Savory)

One of the best small rock-garden or front-of-the-border hostas where full shade is available. Excellent foil for Japanese azaleas. Makes a low-growing, tight mound. Blade lanceolate, tapering to a point, undulate, margin ruffled, of thin substance, $4 \times \frac{3}{4}$ in (10 × 2 cm), pale chartreuse. Slender scape 10 in (25 cm), bearing raceme of mid-lavender flowers. Produces a succession of blooms. Increases rapidly. An excellent hosta.

H. 'Little Aurora' (Aden)

One of the smallest gold-leafed hostas. Suitable for the rock garden, peat bed or as an edging plant. Blade $3 \times 1\frac{1}{2}$ in (9 × 4 cm), rugose, slightly cupped, bright gold, of heavy substance. Scape and raceme almost too large in proportion to the size of the blade and clump. Near-white flowers in midsummer. Grow in semi-shade. Increases more rapidly than many of this type.

H. 'Louisa' (Williams)

One of the best of the many selected seedlings and forms of *H. sieboldii*, and very similar to *H.* 'Emerald Isle'. This was the first hosta ever raised with white-margined leaves and white flowers. Named by Mrs Frances Williams for her daughter. Small, weak-growing, slightly stoloniferous hosta with lanceolate, blunt-tipped leaves up to $4\frac{1}{2} \times 1\frac{1}{2}$ in (11 × 3 cm) which, with slender petioles about twice the length of the blade, make a foliage mound about 12 in (30 cm) high. Clear white trumpet-shaped flowers in late summer. *H.* 'Louisa' only achieves anthesis in optimum cultural conditions. When it does flower, and when well-sited in the garden, *en masse*, it makes a superb display (Pl. 3).

H. 'Love Pat' (Aden)

Seedling of *H. tokudama*. Its raiser thinks it better than the species. Deeper glaucous-blue, leaves more cupped. Leaf shape quite distinct from the better-known *tokudama* seedling, *H.* 'Buckshaw Blue' p. (107).

H. 'May T. Watts' (T & Z)

A white-margined sport of *H.* 'Golden Sunburst' which arose through tissue culture.

H. 'Midas Touch' (Aden)

An *H. tokudama* form. An upright mound up to 14 in (35 cm). Leaves orbicular, cupped and pronouncedly puckered. Bright gold. Pale lavender flowers in midsummer. Withstands full sun in British gardens.

H. montana 'Aureo-marginata'

See *H. montana* f. *aureo-marginata* (p. 70).

H. 'Moonglow' (Anderson)

Depending on growing site, *H.* 'Moonglow' can be brighter and better, although less elegant than *H.* 'Moonlight'. A seedling of *H.* 'August Moon' which occurred naturally rather than from tissue culture. Often, and understandably confused with *H.* 'Moonlight' but leaves more shiny and more cupped. Leaves are bright chartreuse-gold with irregular white margins.

H. 'Moonlight' (Banyai)

A bud-sport of *H.* 'Gold Standard' (Banyai) (p. 112). Slightly less vigorous than *H.* 'Gold Standard'. *H.* 'Moonlight' is a medium to large-sized hosta, and one of the first raised with yellow/gold leaves and a white margin. Blade ovate, cupped and rugose, unfurls mid-green with narrow, clearly defined white margin, then turns bright yellow and finally parchment-white after midsummer if grown in full sun (in southern Britain) to half shade. Needs good cultivation and careful siting to counteract the lack of chlorophyll. An outstandingly attractive cultivar provided the leaves do not become scorched.

H. 'Moorheim' (Royal Moorheim Nurseries)

Noticed by Chris Sanders of Bridgemere Nurseries, Nantwich, Cheshire, growing at the Government Experimental Station at Boscoop, Holland, who had obtained their plant from the Royal Moorheim Nursery, Holland who market it under this name. It appears similar to *H.* 'Antioch'. *H.* 'Goldbrook' and *H.* 'Spinners' but may, of course, have arisen independently.

H. nakaiana (golden form) (Smith) nom. illeg.

It should be known as *H.* 'Birchwood Parky's Gold' (Shaw). It was sent to Eric Smith from the States, who distributed it as *H. nakaiana* 'Golden', thinking it to be a golden form of the type, although slightly larger. Vigorous in spite of its rather fragile appearance and a rapid increaser, forming attractive clumps. Suitable for growing at the edge of light woodland. Some sun is necessary to bring out the gold in the leaves. Blade of thin substance, chartreuse-gold. Good display of mid-lavender flowers on a tall scape. (Fig. 7.)

H. 'Nakaimo' (Imperial Gardens, Japan).

Syn. 'Nakhima'. Introduced from Japan to the USA before 1960. A hybrid of *H. capitata*. According to Hensen (1985) the other parent is likely to come from the section Helipteroides, as it exhibits characteristics typical of that section. Similar in leaf and bract to *H. capitata* but does not set seed.

H. 'Nameoki'

Differing from *H.* 'Nakaimo' (Hensen 1985) in its slightly smaller leaf blade. Of interest today because, like *H.* 'Nakaimo' it illustrates the fact that hostas bearing illegitimate Latinized names were sent from Japan and, until proved otherwise, were usually mistakenly thought to be species. These two hybrids are also of interest as there are very few known *H. capitata* hybrids and they are very much collector's plants.

H. 'Nancy Lindsay' (Grenfell)

(Seems to be synonymous with *H.* 'Windsor Gold'). Thought to be a sport of *H. fortunei* var. 'Hyacinthina'. Named for the late British horticulturalist, Miss Nancy Lindsay, a keen hosta collector, daughter of Nora Lindsay, the well-known garden designer. It was first thought to be a sport of *H. fortunei* 'Albopicta' on account of the yellow base colouring, and was formerly known as *H.f.* 'Aurea Longstock form', as it was erroneously believed to have arisen at Longstock Water Garden (p. 155). Now extensive research shows that this sport actually arose first in Nancy Lindsay's own garden. Medium-sized hosta of stiff, upright habit; an open clump with erect leaves. Blade ovate, incurving; margin flat with faint hyaline line as in *H.f.* 'Hyacinthina', surface mottled gold and green, the gold predominating at first, the colour fading gradually to dull, mid-green after midsummer. Tall, glaucous scape bearing dense raceme of exceptionally attractive deep lavender flowers in midsummer. *H.* 'Nancy Lindsay' associates well with glaucous-leafed hostas but its unusual colouring is not to everyone's taste and is often mistaken for virus infection. There seem to be both a larger and a smaller version of this plant.

H. 'Neat Splash' (Aden)

Small, stoloniferous, ground-covering hosta exhibiting unstable variegation. Blade lanceolate, 5×1 in ($13 \times 2\frac{1}{2}$ cm), streaked and splashed yellow with matt, dark green base. Sometimes the streaking changes into an irregular yellow margin on a green base and then this hosta should be called *H.* 'Neat Splash Rim'. This rule also applies to the slightly larger *H.* 'Yellow Splash' which, if it mutates to a yellow-margined hosta, should then be called *H.* 'Yellow Splash Rim'.

H. 'Nicola' (Eason)

The only green-leafed hosta of the Tardiana group to receive a name and then only on account of its pinkish flowers. Eric Smith made particular note of the fact that they were pink. Blade matt, dark green $6 \times 2\frac{1}{2}$ in (15×6.5 cm). Large raceme of pinkish flowers held well above leaf-mound in late summer. It is grown primarily for its flowers.

H. 'Northern Halo' (Walters)

A sport from *H. sieboldiana* 'Elegans'. The blade is narrower than the type, rugose, glaucous, blue-green with irregular creamy-white margins. Some of the variegation projects inwards overlapping a small band of green, creating an attractive chartreuse effect. The blade is 12×10 in (30×25.5 cm), slightly cupped and held erect, of heavy substance. Similar in flower to *H. sieboldiana* 'Elegans'. It arose in tissue culture and appears to be a good increaser. The creamy-white margin tends to scorch even in a very little sunshine in young plants, but as it matures it becomes a superb hosta. There appears to be a selection with wider margins but this is not visible until the plant is several years old.

H. 'Northern Lights' (Walters Gardens)

An *H. sieboldiana* 'Elegans' sport which arose in tissue culture. The blade is elongate and incurving and slighlty undulate; 10×8 in (25×20 cm), irregularly coloured in the centre with creamy-white and where the variegation overlies the green margin, the colour is chartreuse.

H. 'Northern Mist' (Walters Gardens)

Similar to *H.* 'Northern Lights' but central variegation is mottled cream and green.

H. 'Northern Sunray'

See under *H.* 'George Smith' (p. 110).

H. 'North Hills' (Summers)

Named after the village in Long Island, New York where Alex Summers discovered it. Thought to be a sport of either *H. fortunei* 'Aureo-marginata' or of a white-margined form of *H.f.* 'Stenantha'. Vigorous, fast-increasing, medium to large hosta of stoloniferous habit. Petiole up to 12 in (30 cm) long, angled and deeply grooved and winged. Blade ovate-cordate, shallowly undulate, tapering to a point, slightly rugose, matt, mid-to-dark green with irregular, narrow, clear, sharp white margin, prominent veins. Scape high above leaf-mound, at least 30 in (90 cm). Lavender flowers in late mid-summer. Never sets seed. Covers large areas of ground very rapidly so it is ideal as a landscaping plant. The attractive foliage lasts well until the first frost.

H. 'Oriana' (Smith/Eason)

One of Eric Smith's GL series. An improved form of *H. fortunei* 'Aurea'. Rounder-leafed than most of the other named forms, see *H.* 'Gold Haze, (p. 111).

H. 'Pastures New' (Smith)

Cross between *H. nakaiana* and *H. sieboldiana*.

Fig. 14 *H. plantaginea*
Description on p. 62

Smallish hosta, though larger than *H. nakaiana*. Makes an attractive mound. Blade cordate, margin shallowly undulate, of thin substance, glaucous, grey-green. Slender scape well over-tops leaf-mound. Pale lavender flowers in mid-summer. Makes an excellent tub or specimen plant for a shaded terrace or courtyard. Hostas of similar appearance include *H.* 'Minnie Klopping' (Arett/Klopping), *H.* 'Candy Hearts' (Fisher) and *H.* 'Pearl Lake' (Paine) (American Hosta Society Award of Merit Hosta) with dark purple petioles and bird's head type of raceme. *H.* 'Candy Hearts' is the largest.

H. 'Phyllis Campbell' (Maxted)

A private communication from Eric Smith reveals that he received this hosta from the late Mrs Maxted. It came with a clump of *H. fortunei* 'Albo-picta' and is, presumably, a sport of it. Blade creamy-white, overlaid with fine, dark cross veins, giving a 'netted' appearance. *H.* 'Sharmon' (Donahue) appears to be virtually identical, although it is understood to be a sport of *H.f.* 'Hyacinthina'. *H.* 'Gene Summers' is also similar. All fade to a dull, patchy green towards anthesis but are attractive hostas when they first unfurl and for several weeks afterwards and are well worth growing in a shaded spot.

H. 'Piedmont Gold' (Piedmont Gardens-Payne)

Robust, medium to large clump with leaves having an attractive twist. Blade ovate-cordate, shallowly undulate, margin fluted, 8 × 5 in (20 × 13 cm), slightly rugose, glossy surface, prominently veined, bright gold. White flowers in midsummer. Does best in dappled shade. Sometimes confused with *H.* 'Sun Power' which needs even more shade.

H. 'Po Po' (O'Harra)

Tiny, heart-shaped blue-grey hosta, similar in appearance to the Tardiana group, but a seedling of *H. venusta*. New to the nursery trade. One of the smallest hostas.

H. 'Resonance' (Aden)

Small, ground-covering hosta, similar in its juvenile stage to *H.* 'Ground Master'. As it develops it can be seen that the lanceolate-ovate leaves are much paler yellowish green, more deeply furrowed, and the variegation a deeper yellow. Tall spires of deep purple flowers in late summer. It increases well.

H. 'Royal Standard' (Wayside)

The first and only hosta to be patented. It originated at Wayside Gardens Nursery in the late 1960s and was much sought after when it first reached British nurseries. A hybrid between *H. plantaginea* and probably *H. sieboldii*, like *H.* 'Honeybells', with which it is often confused. The stiffness and slight rugosity of the leaves certainly suggest that *H. sieboldii* was one of the parents. Blade ovate, bright apple-green, appearing chartreuse from a distance when grown *en-masse*. Flowers white, scented, on a tallish raceme. Will tolerate strong sunlight as long as there is adequate moisture at its roots. Flowers from late summer into early autumn and is a good substitute for *H. plantaginea*, a shy-flowerer, when a scented hosta with white flowers is required. A vigorous and rapid increaser, making it an excellent ground cover or landscaping plant.

H. 'Ruffles' (Lehman)

A clump-forming hosta of *H. sieboldiana* origin. One of the first large hostas to have a ruffled margin. Blade up to 12 in (30 cm) long, pointed, mid-green. American hosta hybridizer Dr Kevin Vaughan (*AHS Bulletin*, Spring 1986) states that he is using *H.* 'Donahue's Piecrust' now as a parent for breeding ruffled margined hostas. He finds the species *H. pycnophylla* (pp. 81–2) also produces hostas with heavily ruffled edges. This new development, an advance on the first 'piecrust'-margined cultivars, is producing some very attractive hybrids. Dr. Vaughan is also using *H.* 'Regal Ruffles' (Minks) which has a larger and bluer leaf than *H.* 'Krossa Regal' and has the advantage of being very fertile. *H.* 'Bold Ruffles' is another good one.

H. 'Saishu Jima' (Summers)

Name given by Alex Summers to a hosta received from Japan which was found growing on the island of Saishu. Thought to be identical to *H.* 'Mount Kirishima'. In fact it is a selected form of *H. sieboldii*. A flat-growing, dwarf hosta. Narrow, lanceolate, glossy dark green leaves with undulate margins. An abundance of purple flowers in late summer and early autumn. A vigorous hosta which increases rapidly. Suitable as a foreground plant, for rock crevices or a peat bed.

H. 'Sea Drift' (Seaver)

Mrs Mildred Seaver, Needham Heights, Mass., prefaces most of her cultivar names with 'Sea'. *H.* 'Sea Drift' is another of her excellent hybrids which are so popular with members of the American Hosta Society as is her ruffled margined *H.* 'Sea Octopus'. *H.* 'Sea Drift' has ruffled margins and is an improvement on the original *H.* 'Green Piecrust'. Blade cordate, with a markedly crinkled margin, the surface shiny, prominently veined, 7 × 5 in (17.5 × 12.5 cm). Its heavy substance makes it sun-tolerant and resistant to slug and snail damage. Tall scapes bearing lavender-pink flowers in midsummer. Much used by flower arrangers.

H. 'Sea Lotus Leaf' (Seaver)

A very large, round, puckered blue-green leafed hosta.

H. 'See Saw' (Summers)

A sport of *H. undulata* 'Erromena' (itself a sport of *H. undulata*). An open clump. Blade broad-spatulate, up to 5 × 1½ in (13 × 4 cm), matt, olive-to-dark green, bordered by a narrow, clear white margin which appears too tight for the blade. This effect, which affects many variegated margined hostas, especially those raised by tissue culture, is known as the 'drawstring effect' and was coined by Dr Warren

Pollock in the *AHS Bulletin*, Fall 1988. Tall scape with many leafy bracts and flower bracts which cluster round the raceme. Blooms in early summer. A distinctive and attractive cultivar, especially in the early part of the summer, before the margins become scorched which they easily can if not grown in dappled shade.

H. 'Semperaurea' (Foerster)

A gold-leafed *H. sieboldiana* form grown in East Germany since the 1930s (p. 54). It is thought to have come originally from Japan. (It must have arisen from a gold-margined *H. sieboldiana* pre-dating *H.* 'Frances Williams'). Probably the golden form of *H.* 'Mira'. Leaves larger than those normally associated with *H.* 'Golden Sunburst' (p. 112) and much more glossy. A magnificent hosta when grown to maturity. Needs shade.

H. 'Shade Fanfare' (Aden)

See *H.* 'Flamboyant' (Aden) (p. 109). Medium-sized clump. Blade cordate, flat margin, tip rolled under, 7 × 5 in (18 × 14 cm), matt, light green, bordered by wide creamy-yellow margin. Lavender flowers in midsummer. Very floriferous. An attractive hosta for a shaded border or light woodland, and bleaches if grown in too much sun.

H. 'Shade Master' (Aden)

A seedling of *H.* 'White Vision' × *H.* 'Gold Regal'. Dense mound of mid-gold leaves growing to 22 in (55 cm) tall. Trumpet-shaped, pale mauve, closely-packed flowers produced on 30 in (72 cm) scapes in mid-to-late summer. Vigorous and increases rapidly.

H. 'Shelleys' (Hodgkin/Brickell)

Named by Chris Brickell after the well-known garden of the late British plantsman, Elliot Hodgkin, who collected it in Japan in the 1960s. Graceful, medium-sized clump. Petiole grooved, bright mauve at base. Blade ovate lanceolate, margin distinctly undulate, 9 × 3 in (23 × 8 cm), matt, dark green with veins close together. Scape 10 in (25 cm) long, with occasional foliaceous bracts. Purple flowers in mid-summer. A distinctive and extremely attractive hosta which needs time to establish. Still only a collector's plant but should have a great future as there is nothing else quite like it.

H. 'Shining Tot' (Japan)

One of the smallest hostas offered. Leaves deep green, with a glossy texture and of thick substance. Lilac flowers in midsummer. Will tolerate full sun in Britain, three quarters in the United States. Still rare in the nursery trade.

H. sieboldiana 'Elegans'

See *H. sieboldiana* var. *elegans*, (p. 76).

H. 'Silver Lance' (Savory)

A medium-size hosta with long, graceful dark-green leaves with a thin white margin. Scapes bearing a multitude of delicate lavender flowers in mid-summer.

H. 'Snowden' (Smith)

The best of Eric Smith's crosses between *H. sieboldiana* (white-flowered form) and *H. fortunei* 'Aurea'. Raised at the now defunct Plantsmen Nursery (p. 55). Robust, large, slow-growing hosta eventually making a clump 36 in (115 cm) wide. Blade approximately ovate (nearer in shape to *H. fortunei*), but of heavy substance, rugose, glaucous blue on unfurling but gradually turning glaucous light greyish-green. Stout scape with raceme of green-tinged white bell-shaped flowers. If given optimum cultivation it will grow into a magnificent specimen. However, it is not really commercially viable as it sometimes seems to 'die back' for no accountable reason. Definitely a collector's plant.

H. 'Spinners' (Smith/Chappell)

Distributed originally from Hadspen House House by Eric Smith under the nomen illeg. *H. fortunei aureo/alba* as describing the attractive, irregular creamy-yellow margins which turn white as summer advances, especially if grown in sun. Leaves less cordate than most *H. fortunei* types, margins slightly undulate and graduating to a point. Named after Peter Chappell's nursery garden at Boldre, near Lymington, Hampshire where magnificent clumps grow in perfect semi-woodland conditions (p. 149). Robust, fast-increasing, large hosta, somewhat similar to *H.* 'Antioch' (p. 103), but differing in the poise of the leaves and the greater irregularity of the variegated margin. Must have plenty of shade, and shelter from strong winds. One of the very best of this type. *H.* 'Goldbrook' is similar but noticeably smaller.

H. 'St George's Gold' (Grenfell/Bond, Windsor Great Park)

H. sieboldiana × *H. tokudama* seedling found in the late Mrs Gwendolen Anley's Garden, St George's, Woking, Surrey. Donated to the Savill Garden sometime in the late 1960s or early 1970s. One of the very first of the gold-leafed cultivars, but only recently named and soon to be registered. Of *H. sieboldiana* 'Elegans' size and vigour. Mid-gold leaves assume slight chartreuse tint when grown in shade. Near-white flowers. Associates well with *H. ventricosa* 'Aureo-maculata' and blue *Meconopsis betonicifolia*.

H. 'Subcrocea'

See *H. sieboldii* f. *subcrocea* (p. 97).

H. 'Sugar and Cream' (Zilis)

A *H.* 'Honeybells' sport. Very open mound, leaves arching downwards. Blade similar in shape to *H.* 'Honeybells', mid-green, margin creamy-white with some ruffling. Scape 43 in (120 cm), no leafy bracts, bearing a raceme of near-white to pale lavender flowers in late summer to early autumn.

H. 'Sum & Substance' (Aden)

This hosta has *H. sieboldiana/H. nigrescens* and *H. hypoleuca* in its somewhat complicated lineage. It is undoubtedly one of the best of the very large golden-chartreuse-leafed hostas. A dominant plant in the border. Sun-tolerant even in the southern states of America and almost impervious to slugs and snails. Shiny, dark purple/brown sheaths from which arise 12 in (30 cm) petioles, deeply grooved and dark purple at the base. Blade 14 × 10 in (35 × 25 cm). The leaves have a crumpled appearance as they unroll. Blade cordate-orbicular, slightly concave, margin flat, conspicuously pointed at the tip, of very heavy substance, smooth, matt, the veins widely spaced, unfurls mid, dull green becoming gradually chartreuse to dull, old gold if grown in enough sun, glaucous below. Scape 20 in (50 cm) stout, glaucous, foliaceous bracts near raceme up to 5 in (13 cm) suffused dull purple, flower bracts greyish-purple, purple-edged. Lavender flowers in mid-to-late summer.

H. 'Summer Fragrance' (Vaughan)

H. plantaginea × unknown seedling. Dense, vase-shaped clump. Blade ovate-lanceolate. Both blade and margin undulate, $7\frac{1}{2} \times 3\frac{3}{4}$ in (19 × 10 cm), of thin substance, matt, mid to dark green, bordered with almost uniform, narrow creamy-white margin. Scape up to 42 in (118 cm) without leafy bracts, bearing a raceme of fragrant mauve flowers (flower bud mid-purple). Blooms late summer. Sometimes re-blooms. Fertile.

H. 'Sun Power' (Aden)

Medium-sized, upright, graceful clump with attractive twist to leaves. Blade 10 × $7\frac{1}{2}$ in (25 × 19 cm), shallowly undulating margin, tapering to a point; upper surface light chartreuse to bright gold which is maintained until the first frost, underside distinctly pruinose, veins, close together and prominent. Scape 36 in (115 cm) with leafy bracts, pale lavender flowers densely spaced on the raceme. Blooms in late summer. Some shade is required to prevent the thin-substanced leaves from scorching. Sometimes confused with H. 'Piedmont Gold' (p. 120).

H. 'Sweet Standard' (Zilis)

A H. 'Honeybells' sport. One of the new generation of hybrids from H. 'Honeybells' and H. 'Royal Standard' which are streaked and/or have variegated margins. Blade similar in shape and size to H. 'Honeybells' but streaked green and creamy white. This colouring is variable and creates a somewhat mottled effect. Scape 43 in (120 cm), bearing raceme of pale lavender fragrant flowers in late summer. More a curiosity than a classic hosta but possibly a fore runner in a new line of breeding.

H. 'Sweet Susan' (Williams)

Pod parent H. plantaginea with H. sieboldii as pollen parent. Intermediate between the two. The first hybrid raised with violet, scented flowers. Blooms from the beginning of August to early September. Slightly less fragrant than H. plantaginea. Of historic interest only now as it has been superseded by more garden-worthy hostas.

H. 'Tall Boy' (Savill) (AM)

A hybrid of H. rectifolia, with taller flower scapes and dense racemes of deep blue-purple flowers. Sent to Sir Eric Savill from Gulf Stream Nursery, Virginia, USA in 1961 under the name 'Funkia japonica'. It won an AM for its flowers, but its leaves are disappointing, of a nondescript colour, and with broad, unattractive petioles. It is a classic hosta which should be in every collector's garden.

The Tardiana Group

See pp. 54–8 for historical background. The Tardiana group was raised at a time when to produce a hybrid hosta at all was considered a great achievement and this has given them a particular importance and a certain cachet. What is remarkable about them is that all the plants in the group are derived from a single crossing made once only – Eric Smith's original and daring crossing of H. sieboldiana var. elegans with H. tardiflora, the seedlings of that cross having later been crossed among themselves. They are, therefore, a genetically limited group. It is this that makes it difficult to select the best, though, with the passage of time, some have emerged as continuingly popular while others have been left behind. Therefore, there is no attempt here to describe every one of the group nor, indeed, even every named Tardiana hybrid, but only those which seem to me to be good garden plants. Many nurseries list for sale 'tardiana hybrids'. These are seed-raised forms and therefore not true to type and are likely to have less attractive leaves of a poorer colour. See *H.* 'Halcyon', (p. 114).

H. 'Tenryu'

A giant hosta, possibly a form of *H. sieboldiana* or *H. nigrescens*. Found growing in cultivation in the area of the Tenryu river in Japan. The height of the foliage mound is more than 39 in (100 cm) and the scape is more than 78 in (200 cm) tall. Both surfaces of the leaf are covered in a thick white pruina. Not known in Europe or the United States.

H. 'Thea' (Barcock)

Derivative of *H. fortunei* var. 'Hyacinthina', selected by British plantswoman and hosta collector, Mrs Netta Statham. Named for the late nurseryman Fred Barcock's wife, at whose Drinkstone, Suffolk nursery it arose. Blade ovate, undulate, tapering to a point, 7 × 3 in (18 × 8 cm), glaucous blue-green, irregularly variegated in a creamy yellow, mostly in the centre. Very unstable and not a vigorous hosta and very slow to increase. In some years the variegation is very attractive and others it is not. *H.f.* var. *hyacinthina* has a marked tendency to throw variegated sports and, in consequence, there are many named hostas similar to *H.* 'Thea'. This is inevitable in view of the instability of 'Hyacinthina'.

H. 'The Twister' (Savory)

A seedling or sport from *H. tortifrons*. Small ground-hugging hosta suitable for the rock garden and peat bed. Blade lanceolate, twisted through the centre and along the margin, dark green. There are several similar hostas available now and they do much to dispel the myth that all hostas worth growing are giant-sized plants with glaucous, heart-shaped leaves.

H. 'Tinkerbell' (Williams)

H. sieboldii alba, (probably selfed). This, and *H.* 'Snowflake' (with white flowers) are early hybrids raised by Mrs Williams and although of historic interest only are still sometimes offered for sale.

H. 'Tiny Tears' (Savory)

Seedling of *H. venusta*, but smaller. Blade approximately 1 in (2½ cm) long. Makes a dense mound and is suitable for the rock garden and peat bed.

H. tokudama 'Aureo-nebulosa'

See *H. tokudama* f. *aureonebulosa* (p. 77).

H. tokudama 'Flavo-circinalis'

See *H. tokudama* f. *flavocircinalis* (p. 77).

H. tokudama 'Flavo-planata'

See *H. tokudama* f. *flavoplanata* (p. 77).

H. undulata 'Albo-marginata'

See *H. undulata* var. *albomarginata* (p. 97).

H. undulata 'Erromena'

See *H. undulata* var. *erromena* (p. 101).

H. undulata 'Univittata'

See *H. undulata* var. *univittata* (p. 101).

H. 'Vanilla Cream' (Aden)

Hybrid of *H. tokudama*. Dwarf hosta. Clump 3 in (8 cm) tall × 8 in (20.5 cm) wide, suitable for the rock garden and peat bed. Blade cordate-orbicular, of heavy substance, 2 × 1½ in (5 × 4 cm), light lemon-green to pale chartreuse. Scape 12 in (30 cm), reddish, with some

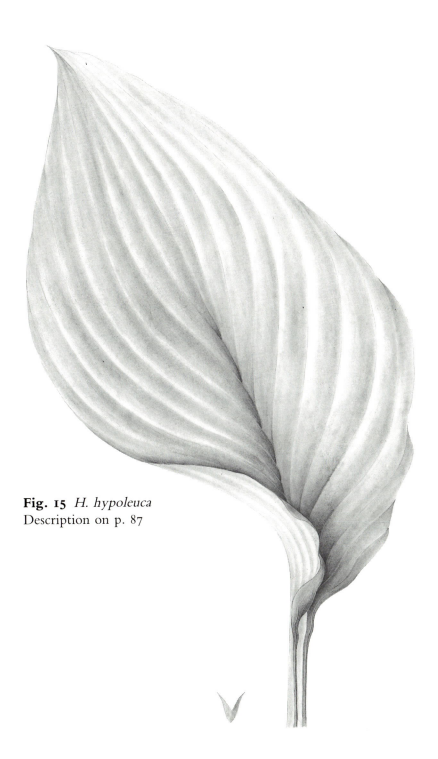

Fig. 15 *H. hypoleuca*
Description on p. 87

leafy bracts, bearing pale lavender flowers in late summer. Needs some shade.

H. ventricosa 'Aureo-maculata'

See *H. ventricosa* f. *aureomaculata* (p. 65).

H. ventricosa 'Variegata'

P. 65. (*H. ventricosa* 'Aureo-marginata' in USA).

H. venusta 'Variegated'

See p. 67.

H. 'Vera Verde' (Klehm)

A strange choice of name as the Latin means 'true green'. Formerly called *H. gracillima* 'Variegated' but since not apparently a sport of *H. gracillima* it was given a cultivar name. The Japanese consider it a dwarf variegated form of *H. cathayana* so until any definite conclusions can be reached it is better that it has a cultivar name as it is now in the nursery trade. A small, very stoloniferous hosta suitable for the rock garden and peat bed. Petiole 5 in (12.5 cm) long, shallowly grooved at the base, almost flat near the junction with the blade, fine white margin. Blade decurrent, lanceolate, blunt tipped, slightly rugose, matt olive-green with slightly undulate thin creamy-white margin. Scape 6 in (15 cm), thin, pale olive-green with one or more variegated-margined foliaceous bracts near raceme. Pale mauve flowers in mid-to-late summer. A very distinctive little hosta.

H. 'Wagon Wheels' (Minks)

Similar to *H. tokudama* 'Flavo-circinalis' but the blade is rounder. Very slow to increase.

H. 'Weihenstephan' (Mussel)

A selected form of *H. sieboldii alba*. Raised at the Trial Garden of the Weihenstephan Brewing & Horticultural Faculty at Friesing, Munich, and named by its director, Herr Mussel. Blades wider and more rugose than the type, elm-green. The erect scapes carry a profusion of white trumpet-shaped flowers, larger and more prolific than the type. At its best in a massed planting at the front of a border backed by dark foliage.

H. 'White Christmas' (Krossa)

A probable sport of *H. undulata*. Small, clump-forming hosta with undulate, spatulate leaves with white predominating, and irregular green margins. Very slow-growing and unfurls late in the season after most other hostas. Showy and attractive if it can be cossetted into performing well. Best in shade.

H. 'Wide Brim' (Aden)

Showy, clump-forming hosta, fairly similar to *H. opipara*. Blade cordate-orbicular, margin flat, rugose, of heavy substance, 8 × 6 in (20.5 × 15 cm), glaucous dark blue-green with wide, creamy-white irregular margin. Some forms of this hosta may not have such rugose leaves. An attractive new hosta with great potential for flower arranging.

H. 'Wogon Gold' (Krossa)

A gold-leafed form of *H. sieboldii*, differing from *H.* 'Subcrocea' by its wider and stubbier leaf blades, from the island of Honshu. Has lovely rich gold leaves and fine purple flowers on tall scapes, followed by purple seed pods. Of stoloniferous habit. Needs shade. It was introduced to Britain by Dr Rokujo of Tokyo University, who called it 'Wogon Giboshi', but in America it has always been known as 'Wogon Gold'. It might be simplest to call it 'Wogon' which means golden in Japanese. Dr Rokujo introduced many hostas to Britain in the 1960s, mostly direct to the Royal Horticultural Society's Garden at Wisley or to Hydon Nursery, at Godalming, Surrey.

H. 'Zounds' (Aden)

One of the larger gold-leafed hostas of the *H. sieboldiana* type. Leaves appear crumpled on unfurling. Blades slightly dished, very rugose, of heavy texture, glossy, bright gold with a metallic sheen. Pale lavender flowers in early summer. Useful for lighting up dark corners of the garden. (This comment is more applicable to the heavily shaded gardens of America than to British gardens).

5.

USES OF HOSTAS

Hostas for flower arranging

Flower arrangers were among the first people to take a real interest in hostas; their use of hostas did more than anything else to popularize the genus. When Constance Spry was at the peak of her fame in the 1950s, the range of hostas was very limited, four or five varieties at the most, whereas now there are at least 100 types suitable for flower arranging (some brought to Britain by Sheila MacQueen who recognized their potential, long before the nursery trade had ever heard of them). George Smith, who for many years had been growing a wide variety of suitable hostas in his Yorkshire manor-house garden, has done much to promote their use as flower-arrangement material in his colourful demonstrations all over the world. He even has a hosta named after him (p. 110). But he still chooses *H. fortunei* var. *aureo marginata*, our old friend 'Yellow Edge', as the hosta he would never be without.

Many of the members of the British Hosta and Hemerocallis Society are flower arrangers and the great surge of interest in hostas in the 1960s and 1970s on the part of nurserymen was mainly a result of the flower arrangers' quest for ever newer and more interesting foliage material.

Between 1954 and 1959 the flower arranging movement was organized under the aegis of the Flower Committee of the Royal Horticultural Society and 1955 saw the first flower-arrangement tent at the Chelsea Flower Show. The movement was naturally fairly dormant during the war years although floral art had been becoming more popular both as a profession and a hobby since the 1920s and 1930s. Its origins perhaps go back to Gertrude Jekyll who published her book *Flower Decoration in the House* as early as 1907; as well as being a plantswoman and garden designer she was obviously no mean flower arranger.

After the end of World War II Julia Clements VMH was one of the first of the leading flower arrangers to recognize the need for flowers in our lives once again, and her lectures and demonstrations started to put flower arranging back on the map. She often used simple pieces of wood and foliage to augment the flowers which were then still in short supply. The upsurge of interest in the decorative arts both in the home and in the garden

Plate 1   Engelbert Kaempfer's drawing no. 166
Reproduced by courtesy of the British Library

Plate 2   Engelbert Kaempfer's drawing no. 52

Reproduced by courtesy of the British Library

Plate 3 A superb form of *H. sieboldiana* 'Elegans' growing at Hollybank House, Woking

Plate 4   *H. crispula* exhibiting its characteristically twisted leaves. Growing in the
garden of Doreen and Geoffrey Stevens, London

Plate 5   *H. kikutii* var. *pruinosa* exhibiting its beak-like flower-buds
Reproduced by kind permission of George Schmid

Plate 6   Dense flowerhead of *H. kikutii*
Reproduced by kind permission of George Schmid

Plate 7 (top)   *H. opipara*, a hosta new to cultivation in the West. The cream margin would be wider in a mature plant
Reproduced by kind permission of Hajime Sugita

Plate 8 (left)   *H. kiyosumiensis* growing on the bank of a stream at Nukata, Japan
Reproduced by kind permission of Hajime Sugita

Plate 9 (above)   *H. longipes* growing on a mountain side at Shidara
Reproduced by kind permission of Hajime Sugita

Plate 10 *H. undulata* 'Univittata', a hosta with unstable variegation

Plate 11 (top)   A young plant of *H. yingerii,* one of the newly published species,
collected in Korea
Reproduced by kind permission of George Schmid

Plate 12 (bottom)   A narrow-leafed form of *H. sieboldii* 'Alba' growing in typical Japanese fashion
Reproduced by kind permission of Hajime Sugita

Plate 13 (top)   *H. ventricosa* 'Aureo-maculata', showing its brilliant-yellow variegation
in early summer

Plate 14 (bottom)   *H. montana* 'Aureo-marginata' growing in perfect contrast with
H. sieboldiana in the Hosta Walk at Apple Court

Plate 15   H. fortunei 'Aurea' in early summer

Plate 16   H. 'Julie Morss', a rare and beautiful cultivar

Plate 17   *H.* 'George Smith' growing at The Manor House, Heslington; the home
of George Smith, International Flower Arranger

Plate 18   H. 'Frances Williams', sumptuous in flower as well as in leaf
Reproduced by kind permission of Julian Grenfell

Plate 19 (above) H. 'See Saw' growing in the Hosta Walk at Apple Court, part of the National Reference Collection

Plate 20 (below) H. 'Gold Standard', one of the most popular cultivars raised in the last 20 years

Plate 21 (left) H. 'Buckshaw Blue', photographed at Gene and Alex Summers' garden, Bridgeville, Delaware, during the 1987 American Hosta Society's Convention Tour, when it was awarded the Nancy Minks Award for the best, registered, small-to-medium hosta seen on the Tour

Plate 22 (right) H. 'King Michael', a sumptuous specimen hosta photographed in the Bridgeville, Delaware garden of Gene and Alex Summers. An improved form of *H.* 'Green Acres'

Plate 23 (bottom) H. 'Ground Master' growing near the Alpine House at the Royal Horticultural Society's garden, Wisley

Plate 24 H. 'Nancy Lindsay' growing amongst hostas of the Tardiana group in the
National Reference Collection at Apple Court

Plate 25 H. 'Golden Tiara', a beautiful clump-forming hosta growing in the garden
of Doreen and Geoffrey Stevens, London

Plate 26 (top)   *H. sieboldiana*, in total contrast to the formally clipped tsuga at the
Ladew, Maryland, topiary garden

Plate 27 (bottom)   *H. sieboldiana* 'Elegans' growing with *Picea pungens* 'Glauca' and
Cupressus glabra, a study in blue, contrasting with *Milium effusum* 'Aureum'

Plate 28 (top)   H. 'Semperaurea' growing in a shade house at Weihenstephan, Munich

Plate 29 (bottom)   The pruinose, purple, sprouting shoots of a hosta of the Tardiana group

Plate 30   An exhibit of *Hosta* mounted by the Royal Horticultural Society at
The Chelsea Flower Show in 1969

Photographed by the late Eric Smith and reproduced by kind permission of Ronald Smith

Plate 31   H. 'Louisa' growing with *Acer palmatum* 'Aureum' at Bressingham Gardens

Plate 32   H. fortunei 'Aureo-marginata', an excellent border plant, grown to perfection
at Bressingham Gardens

Plate 33   A typical hosta planting at Spinners, Hampshire, showing *H.* 'Spinners'
amongst hostas of the Tardiana group

Plate 34   *H. undulata* 'Erromena', well-suited to its waterside situation at
Longstock Water Garden, Hampshire
Reproduced by kind permission of Dick Kitchingman

Plate 35   *H. sieboldiana* seen at its best planted in broad sweeps at
Longstock Water Garden, Hampshire

Plate 36   The golden hosta border in the woodland garden of Ali and Warren Pollock,
Wilmington, Delaware
Reproduced by kind permission of Ali and Warren Pollock

Plate 37 Blue meconopsis planted as a contrast to *H. fortunei* 'Aurea' at
The Savill Garden, Windsor Great Park
Reproduced by kind permission of Lyn Randall

Plate 38 (left)   *H. fortunei* 'Hyacinthina' at Greyrigg, Cumbria with a large planting
of *H. lancifolia* on the rock above
Reproduced by kind permission of Dilys Davies

Plate 39 (above)   The Hosta Walk at Hadspen House, Somerset, planted by
the late Eric Smith

Plate 40 The Hosta Walk at The Walled Garden, Claremont, Surrey
Reproduced by kind permission of Mrs Cecily Martin

Plate 41   A young plant of *H.* 'Sum and Substance' growing in the Hosta Walk at Apple Court

Plate 42   *H.* 'Wide Brim', a modern cultivar with beautifully textured leaves.
Very suitable for flower arrangements

Plate 43   A mature clump of *H.* 'Halcyon', producing a variegated sport
Photographed at Heslington Manor and reproduced by kind permission of George Smith

Plate 44   Part of the hillside planting of hostas grown for flower arrangement
at Vicar's Mead, East Budleigh, Devon
Reproduced by kind permission of Elizabeth and Harold Read

Plate 45   Part of the National Collection of large-leafed species and cultivars of *Hosta*
at Golden Acre Park, Leeds Parks Department, Leeds, Yorkshire
Reproduced by kind permission of Terry Exley

Plate 46   A flower arrangement showing *H.* 'George Smith', by George Smith

Reproduced by kind permission of Webb & Bower

Plate 47 (above)   Mature clumps of *H. sieboldii* at The Savill Garden, Windsor Great Park

Plate 48 (below)   The south end of the newly planted Hosta Walk at Apple Court

Plate 49   H. hypoleuca, painted by Jenny Brasier
and approximately 60% life-size

at the time was probably an antidote to those years of austerity when plants were grown only to eat, and signs of embellishment to our homes were considered a mark of frivolity.

Lack of staff, for one reason or another, after the war, led to the abandonment of the traditional herbaceous border in many gardens and indeed any form of labour-intensive decorative gardening, which led in turn to an increase in interest in shrubs and foliage plants.

In spite of there being such a limited range of hostas available in those early days, what few there were were put to such good use by flower arrangers that nurserymen and hybridizers were quick to see their potential, and even produced hosta leaves with multi-variegation by colchicine treatment and X-rays. However, flower arrangements good enough for royal weddings at Westminster Abbey and other prestigious events were enhanced by the use of hosta leaves of even the simplest form. The only hostas grown in gardens then were the large lance-leafed, sage-green *H. fortunei* (exact variety unknown) with a long pointed tip and undulate margin. White-margined *H.* 'Thomas Hogg', invariably called *H. crispula* (the much rarer and more desirable species), and the green-and-white *H. undulata* forms, all added a touch of coolness and freshness to pedestal and table arrangements. *H. sieboldiana* provided the dense, round shapes and glaucous leaf mass so important for pedestals and large exhibition work. *H. fortunei* 'Albo-picta' and its form, *H.f.* 'Aurea', were useful early in the summer before the variegation and gold colour faded, although 'Aurea', adored by flower arrangers, lacked substance and, being soft, was difficult to use successfully. According to the late Sybil Emberton in her book *Garden Foliage for Flower Arrangement*, Faber, 1968, these two hostas were then considered to be too short in the stalk. All these hostas are still being used today for flower arranging, apart from *H.f.* 'Aurea'. *H.f.* 'Albo-picta' has also probably been superseded by *H.* 'Elizabeth Campbell' (p. 108) and *H.* 'Gold Standard' (p. 112), and there are now many permanently gold-coloured leaves to choose from. They range in leaf size from the tiny *H.* 'Little Aurora' to the majestic *H.* 'Gold Regal'.

In order to fill the flower arranger's palette, the choice of suitable hostas has to be carefully made, especially if they are to be grown in a small garden, say of less than a quarter of an acre (0.1 hectares). Restraint must, therefore, be exercised in choosing those hostas which not only enhance the garden without overcrowding it, but also offer the best in leaf shape, colour, texture, substance, line and variegation. To give an example of how to grow several hostas in a smallish space, we start with one of the larger-leafed hostas, glaucous, grey-blue *H.* 'Krossa Regal', with its undulate margin and upturned tip, which is a first choice on several counts. The clump is urn-shaped with plenty of room under the leaves, in front of its upward and outward-arching petioles, in which to grow *H. undulata* var. *univittata* and the bright gold, corrugated-leafed *H.* 'Midas Touch': vivid contrast of colour and form. Or try echoing the shape of *H.* 'Krossa Regal' with *H.* 'Golden Prayers'. There are many such combinations which would both enhance the garden and an arrangement and not take up too much growing space.

Although hostas with pronounced ruffled margins are becoming popular, it is better to choose one good variety such as *H.* 'Regal Ruffles', *H.* 'Bold Ruffles' or *H.* 'Sea Drift' which will make a definite impact in the garden or in an arrangement, rather than several, which would be overwhelming. There is in the end nothing to beat clean, simple lines. Two outstandingly good medium-sized hostas with wide cream margins and leaves of good substance and texture are *H.* 'Shade Fanfare' and *H.* 'Wide Brim'. They are still very hard to come by, but well worth waiting for. If interesting venation is wanted, then *H.* 'Green Acres' and indeed, all other forms of *H. montana* and its hybrids would do very well. *H.* 'Green Fountain', a hosta with lance-shaped, glossy leaves, still fairly new, is already finding favour with those who seek plain green leaves of good poise and line.

It is perhaps not a good idea for the flower arranger with a small garden to grow hostas with unstable variegations, which can look different every year.

Leaf surface is very important for arrangements and here hostas have a great deal to offer. The most puckered leaves of all are found in the *H. sieboldiana*s, *H. tokudama*s and their hybrids, one of the best being *H.* 'Frances Williams', but these need careful placing in the garden, in the shade but away from any falling detritus from trees which can get trapped in the deep surface indentations.

Yellow-and-green leaves and white-and-green leaves are better not used in the same arrangement. Glaucous-leafed hostas associate best with white variegation, and gold or green leaves with yellow or cream variegation. Bright gold with deep glaucous blue can often be too much of a contrast. This maxim applies just as much to growing hostas in the garden.

Conditioning

Weather conditions and the type of hosta will dictate the time needed for conditioning and also the end result. The material should be gathered early in the morning or in the evening when transpiration is at its lowest – very important in the case of hostas with a large leaf area. The leaves should be left in a cool dark place until needed, if possible for about 12 hours. It is better to pick the leaves earlier than needed and leave them standing in water for a longer time than to rush the period of conditioning. They should be cut from the plant with a clean, sharp knife or scissors, the stems on a slant to increase the area available for water absorption. This one-sided arrow-shape will also pierce the flower-arranger's foam more readily, making placement easy and, again, increasing the area available for water absorption. The leaves and/or flowers should be placed in water immediately because if the ends are allowed to dry out, the intake of water will be restricted and conditioning will be impaired.

This is all the treatment mature hosta leaves and flowers need, but very young foliage may need further treatment. If the leaves feel soft and sappy then place the final 1 in (2.5 cm) of the stem into boiling water, count up to ten, then submerge the whole leaf in a bowl of tepid water for up to two hours, but no more. They can then be placed with their stems in a bucket of water and left to soak with the mature leaves.

Nothing further need be done and they can be placed one on top of another, like stacked chairs, in a plastic bag, making a compact parcel for transportation. With each supporting the other, there should be no fear of broken or damaged leaves.

Hosta leaves do not generally dry very successfully but the seed-heads and sere flower stalks make excellent winter arrangements. They should be left on the plant for as long as possible but removed well before the autumn gales which will scatter the seeds. Judgement and a good long-term weather forecast are needed. The stalks and seed heads can be varnished but many arrangers prefer to use them in their natural state.

Hostas suitable for flower arranging

GREEN LEAVES

H. 'Green Acres'
   'Royal Standard'
   'Green Fountain'
   montana
   elata
   ventricosa
   minor
   tortifrons
   tardiflora

GOLD/YELLOW LEAVES

H. 'Golden Prayers'
   'Gold Edger'
   'Midas Touch'
   'Little Aurora'
   'Zounds'

GLAUCOUS-GREY LEAVES

H. 'Krossa Regal'
   'Snowden'

GLAUCOUS-BLUE LEAVES

H. 'Blue Moon'
   'Blue Wedgwood'
   'Halcyon'
   sieboldiana var. *elegans*
   'Big Daddy'
   'Bold Ruffles'
   'Regal Ruffles'
   'Sea Drift'

GREEN/YELLOW VARIEGATED LEAVES

H. fortunei var. *aureo marginata*
   'Shade Fanfare'
   ventricosa 'Variegata'
   'Elizabeth Campbell'
   'Gold Standard'
   'Golden Tiara'
   fluctuans 'Variegated'

WHITE/GREEN VARIEGATED LEAVES

H. 'Thomas Hogg'
   'Goldbrook'
   undulata & hybrids
   'Francee'
   'Gloriosa'
   'Ginko Craig'
   crispula

YELLOW/GLAUCOUS-BLUE LEAVES

H. 'Frances Williams'
   'Wagon Wheels'
   tokudama forma 'Flavo-circinalis'

FLOWERS

H. ventricosa
   plantaginea
   'Gold Regal'
   'Frances Williams'
   'Halcyon'

Hostas for exhibiting

There are several different ways in which hostas are used at flower shows or as decorative material on exhibition stands, and each requires different methods of preparation.

In 1968 the Royal Horticultural Society's Scientific Section mounted the first exhibit of hostas ever staged in Britain and this in no small way helped to publicize the genus, still then virtually unknown to most gardeners. Since then no other exhibit purely of hostas had been mounted at Chelsea until May 1988 when Sandra Bond, Goldbrook Plants, created an educational exhibit of hostas for which she was awarded a Gold Medal. Her hosta exhibits had twice before been awarded Gold Medals at the Royal Horticultural Society's Summer Flower Shows.

Hostas used purely as background furnishing material, or as a foil to the vividness of azaleas and rhododendrons, have been a feature of plant exhibits for many years. When used in this way large clumps are lifted two weeks before the start of the show, wrapped in wet hessian (burlap), watered regularly and kept under cover. Treated like this they remain in pristine condition for several weeks.

When mounting an exhibit using hostas in pots, things are not nearly so simple, and much early preparation has to be done to ensure that the plants live up to their reputation as foliage plants of the highest rank.

Most flower shows in Britain at which hostas are exhibited are held between May and July when the leaves are at their best and many are likely to be in flower. The hostas should be potted up in July of the year prior to the show, preferably in plastic pots to retain moisture, allowing plenty of room between the clump and the sides of the pot for further root growth. A peat-based coarse compost, with grit, should be used, of medium-to-acid pH. A well-known exhibitor in the USA uses 1 part Pro-Mix, 1 part rotted cow manure, 1 part loam, $\frac{1}{2}$ part sand, and $\frac{1}{2}$ part Perlite, to which is added 1 4 in (10 cm) pot of superphosphate per bushel; different climatic conditions would account for the choice of a richer mixture. In both instances slow-release pellets of 'Osmocote' fertilizer are added to the pots as well as slug pellets. Although it is best to root-prune in spring, if the hostas are potted in July for exhibiting the following year, then it must be done as they go into the pots, but it is more usually done in spring when moving the plants to pots one size up to give room for the addition of new compost. Root-pruning is simply a matter of pruning long, straggly roots back to the main arteries to encourage them to produce new vigorous feeding roots high up in the compost rather than at the bottom of the pot. This should be done with a sterile sharp knife and the wound dressed with Benlate to prevent infection.

The pots should then go into Rokolene shade tunnels on a base of black polythene topped with pea-grit to inhibit slug and snail activity. At the beginning of December, when the leaves have died down, the pots should be transferred to a polythene tunnel and kept fairly dry for the winter, the dead leaves having been removed. In parts of the United States, where the temperatures can be much lower than in Britain, hostas in pots are kept at $55°F$ $(13°C)$ to start with and at $45°F$ $(7°C)$ if the pots have already frozen, so that they do not thaw out too quickly. Experimentation is necessary to take into account climatic conditions and the time needed to produce hostas at their best. In Britain the use of heat is not normally required while the hostas are dormant. Some degree of heat would be needed in northern continental Europe.

As the leaves start into growth in late winter, shading from the sun is needed in order to prevent leaf scorch. In Britain the pots are moved to a 40 per cent-shade tunnel as soon as the leaves appear, so that leaf characteristics can develop. *H. montana* 'Aureo Marginata' is the first to sprout, followed by *H. lancifolia* and *H. tardiflora*, then *H. ventricosa* and its forms, *H. undulata* and its forms, *H. plantaginea*, *H. nigrescens*, *H. crispula* and *H. decorata* and, last of all the *H. sieboldiana* and its forms and *H. tokudama* and its forms. Apart from *H. montana* 'Aureo marginata', this is certainly not the same order in which they sprout when grown in the ground: *H. crispula* is then always one of the earliest.

Hostas with blue foliage colour better in shade and can be placed under the leaves of larger hostas so that less light penetrates. Some hostas with gold foliage colour well in a net tunnel but others need more sun to bring up the colour and should be placed outside, preferably in a sheltered area facing south or south-west. These include *H.* 'Golden Medallion', *H.* 'Golden Prayers', *H.* 'Gold Drop', *H.* 'Midas Touch' and *H.* 'Sum and

Substance'. They must be watched carefully to ensure that they do not burn. The tunnel should be dusted with HCH to prevent woodlice which can chew the petioles. If there is high insect activity then the tunnel is given a smoke of Lindane.

Watering should be started again as the leaves first appear. The pots should be soaked every two/three days, not daily as plants need oxygen as well as water, and this is better than a daily sprinkle unless the air temperature gets too high or lacks humidity; then more watering is needed. Hostas should not be watered from overhead, but directly into the pot to prevent leaf scorch and marking from lime deposits. The glaucous leaves suffer the most from water marks.

If the season is late and the hostas are needed for an early show (e.g. Chelsea), the *H. sieboldiana*s, *H. tokudama*s, and *H. ventricosa*s and related hybrids have to be given a little help to get going and should be placed on a heated bench for a few days, and then left in an unheated greenhouse, and moved from there to a net tunnel. If minimal heat is given then there is no leaf distortion or softness and a 'forced' appearance is avoided. Only if frost threatens do they return to the polythene tunnel.

Plants on exhibition stands change rapidly if conditions are adverse. The light levels may not be high enough to maintain true gold colouring for more than two or three days and all the plants will lose intensity of colour; the greens will fade, the blues can darken. It is essential to have enough plants prepared to make substitutes when necessary.

Experience has shown that hostas have extremely high degrees of tolerance to sun and heat. They have withstood temperatures of $100°F$ ($38°C$), showing no signs of stress after two days, even though the night-time temperature did not drop below $90°F$ ($32°C$). They were kept well-watered and syringed throughout the day with a fine spray of tepid water. To keep the leaves turgid in extreme heat the pots can be arranged on the stand with spaces between the pots filled with balls of wet newspaper and the whole thing topped with moss which is then saturated and kept permanently wet. The moss should be watered with Slugit to kill off any occupants.

After a show the hostas need to be rested. It takes about a month for the true colour to come back and for new leaves to grow to cover damaged or dirty ones, although after June there is no real new growth.

The majority of hostas do very well in pots, especially the *H. fortunei*s and related hybrids, e.g. *H.* 'Francee'; also *H.* 'Golden Tiara', *H.* 'Zounds', *H.* 'Sun Power', *H. minor*, *H. venusta*, *H.* 'Snowden', the Tardiana group, *H.* 'Krossa Regal', the *H. montana*s, *H.* 'Golden Prayers', the *H. nakaiana* types, *H.* 'Hydon Sunset', *H. lancifolia* and *H. sieboldii*. The least successful and the most demanding to prepare are the *H. sieboldiana*s, especially *H.* 'Frances Williams', but *H. tokudama* forma *flavo circinalis* is fine. Other less succesful sorts are *H. ventricosa*, *H. plantaginea* and its hybrids, *H.* 'Midas Touch', *H.* 'Piedmont Gold' and *H. elata*. The *H. fortunei*s actually seem to grow better in pots than in the ground, probably responding well to the very high degree of cosseting.

Pots, tubs and containers

In their native Japan, hostas are usually grown as pot plants simply because gardens are so small or virtually non-existent; sometimes they are even taken inside the house and placed beside an open screen. In Europe they are not usually satisfactory as house plants for more than a week or two because the atmospheric humidity is much lower than in Japan.

Hostas do, however, make sumptuous and striking plant material for growing outdoors in containers, whether of wood, stone, plastic, fibreglass, lead, terracotta, brick or concrete; from old chimney pots to antique stone or lead urns; old municipal granite or stone sinks and troughs or even large square concrete planters outside modern office buildings. The only type of container unsuitable for growing hostas are the very shallow, antique stone sinks.

There is one important proviso concerning hostas grown in containers. It is not advisable to keep the same hosta in a container for more than four years, which is as long as it takes for a hosta to reach maturity. After that, even if well-fed and kept moist, the root system will outgrow its confined space and the hosta will start to deteriorate. It is best to start again.

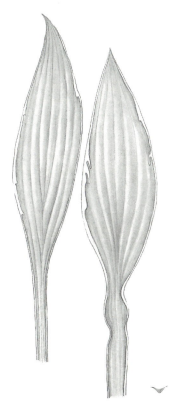

Fig. 16 *H. sieboldii*
Description on p. 96

Many hostas have a very large leaf surface which loses moisture rapidly and this can cause wilting and browning at the edges, unless there is adequate moisture at their roots, supplemented by foliar feeding at half-strength every week during the growing season. In very hot, dry weather it is also a good idea to spray the leaves with water early in the morning. The containers should be placed out of the direct rays of the sun, or should have sun on them for only a short time; they should be protected from exposure to wind and placed well away from draughty corners, since wind can bruise or even split the leaves.

The containers should be raised off the ground with bricks to allow the free passage of air and drainage and to keep the container clean and dry, but a watch should be kept for slugs and snails which can lurk underneath. Before planting, a layer of crocks made from broken terracotta pots should be placed on the bottom of the container covering the drainage hole. The container should then be filled to within 2 in (5 cm) of the top with a rich mixture of potting compost. For a large tub that is to contain a large hosta, a layer of animal manure near the bottom will give extra nourishment. Grit to assist drainage should be mixed with the potting compost as well as charcoal to keep the soil sweet. In each succeeding year a handful of 6X (reconstituted chicken manure) should be placed round the hosta (taking care not to let it come into contact with the petioles or leaves) fairly soon after the crowns appear in order to give the leaves a nitrogen boost. The foliar feed should, however, contain a balanced mixture of nutrients.

Art as well as science and common sense is involved in choosing which hostas to grow in containers. If the area where the hostas are to grow is very small – a tiny courtyard for example – with narrow gaps between the pots and tubs, then obviously hostas with long, splayed or drooping leaves (*H. crispula*) would be unsuitable. The leaf tips can easily be damaged if knocked against and will then go brown. Far better is a smallish hosta with stiff, upright leaves, such as *H. tokudama* forma *aureo-nebulosa*.

Quite a number of hostas change their leaf colour quite considerably during the early part of the summer; even the glaucous-blue types appear almost light green when they first unfurl their leaves, and the majority of the gold-leafed sorts are not gold at all early in the summer – so make sure that any colour schemes blend with the colour that the hosta leaves will finally attain. Consequently, it is probably not a good idea to use hostas such as *H.* 'Janet', 'Gold Standard', 'Moonlight' or *H. fortunei* 'Albo-picta' in containers at all, since their leaf colour changes could upset the balance of a carefully planned colour scheme.

Hostas can be very colourful as foliage plants so care should be taken to make sure that they do not clash with the surrounding walls, paving and garden furniture. A large, halved, unpainted beer or wine barrel containing a huge clump of *H.* 'Frances Williams' will look stunning if placed against a grey stone wall in deep shade, or in front of a natural or creosoted fence or outbuilding; but if placed in front of a white or red brick wall, it will just look garish. 'Frances Williams' is visually a very powerful plant when mature, and so needs careful placing. The adjacent planting should be of delicate ferns to balance its strong texture, colour and bulk. Plain green-leafed hostas, or those with green and white variegated leaves look best against terracotta or brick and will, in fact, look well against most colours. Very dark green leaves (*H. ventricosa*) complement yellow brick and so even more do those with dark green leaves and yellow or gold margins that tone with the brick. Hostas with yellow, gold and cream variegated leaves look good against shiny, white paint but some of the brighter-toned, golden-leafed hostas do not look their best against some of the mid-red bricks. Purplish brick is a very good background to most hosta leaves other than the brilliant golds. Be careful, too, of growing some of the brighter golds in confined spaces. These hostas look best in a modern setting with spiky, architectural plants such as *Phormium* 'Radiance' and *Helleborus corsicus*, planted in squares let into paving slabs on a wide terrace.

Fortunately many gold-leafed hostas have pale mauve, milky-white or even dead-white flowers (*H.* 'Piedmont Gold'), which are quite innocuous, often pleasing and which sometimes even enhance the leaves; but there are some gold-leafed sorts which have bright mauve or purple flowers. These can cause a strident clash in a scheme which has containers placed in close proximity to one another, in which case it is probably better to remove the offending flowers (the leaves will grow bigger) as they will add nothing to the composition. This clash of flower and leaf is not nearly so noticeable or offensive in a landscape or border context.

Many gardeners quite naturally wish their terraces to be sun-traps but should be aware that the reflected heat from the paving or slabs can be punishing to large-leafed hostas, especially those in containers; these should be in dappled shade for at least half the day, or certainly when the sun is at its hottest.

The round leaves of *H. sieboldiana* and *H. tokudama* types always look well in round containers. In a quiet corner *H.* 'Green Fountain' cascading out of a tall, cylindrical terracotta chimney-pot would add charm and repose to a shaded courtyard or terrace. The hosta should actually be growing in a pot, the rim of which would be lodged on the lip of the chimney pot. Variegated ivies are an excellent foil to hostas grown in this way.

In a large tub, say one 3 ft (90 cm) across, *H. undulata* var. *erromena* can be the central pivot to a circular fringe of *H.* 'Ginko Craig', or *H.* 'Francee' with *H.* 'Gold Edger'. The possibilities are many, but try to use together hostas with the same rate of growth.

If in doubt it is always safest to go for a plain green-leafed hosta with shapely leaves; one could hardly go wrong with *H. ventricosa* which also has extremely attractive flowers.

How and where to position the hosta containers is a matter of personal taste and the dictates of the site in question, but a pair, or even four, matching containers of the same sort, planted identically, have more impact than just one container, or four assorted containers with varying ranges of plant material. Accentuate a flight of stone or brick steps with a pair of *H.*

'Krossa Regal' to the side of each top step, or the four corners of a long lily pool with stone urns filled with *H. undulata* or *H. tardiflora*, a leathery-leafed hosta whose surface would echo that of the pads of the water lily. No more than two different materials for the container and the hard landscaping should be used together: for example, stone and terracotta, concrete and brick, lead and York stone. As with the planting, simplicity is all.

Courtyards and town gardens

Courtyards, unlike terraces, are usually enclosed by walls or fences which often cast shade throughout the day, creating the sort of cool, moist conditions in which hostas can flourish. Many small town gardens are simply larger courtyards, enclosed by buildings which also create similar cool, moist growing conditions. Both can make perfect places for the cultivation of hostas, whether in large containers, in borders or in gaps left in the paving. In these conditions hostas will, however, produce fewer flowers and less dense clumps and their leaves will be thinner than when grown in sunnier conditions.

Hostas and other perennials will look most at home, and will perform better, if bare walls or fences are broken up by plantings of shrubs with strong foliage, the shrubs creating sheltered bays or niches for the hostas. This sort of planting could include fatsias (both plain-leafed and variegated) which look good in the angle between two walls, mahonias, bamboos, hydrangeas, maples and ivies and, if the soil is acid, some rhododendrons and evergreen azaleas. Herbaceous plants which grow well in shaded courtyards are ferns, astilbes, bergenias (better separated from the hostas), polygonatum, smilacina, geraniums and kirengeshoma (if the soil is moist enough). Hostas associate well with all these plants, except the bergenias, which although invaluable near paving, have leaves too close in shape to many hostas for pleasing plant association.

A courtyard planting using only Japanese plants, furnished with Japanese stone lanterns and stone water bowls, gravel and sand, is an attractive way of displaying hostas and is quite in keeping with a shaded site.

Hostas often have most impact when they are the only form of plant material in a courtyard and can interact with other components – the paving, walls, pergola, seating, water. At the de Belders estate, Kalmtout, in Belgium, a single specimen of *H. montana*, waist-high and 4 ft (1.20 m) across, dominates a small, enclosed courtyard space. Few gardeners have the discipline or foresight to resist some accompanying planting.

Another splendid piece of restrained planting occurs at the main entrance to Dumbarton Oaks, the enchanting garden in Washington, DC designed by Beatrix Farrand for Mildred Bliss. One enters from the stone pavement of R Street through a black iron gate set in a high brick wall, and in the angle between the wall and the porter's lodge is a small bed planted only with *H. ventricosa*. Not only does this hosta have leaves of superb shape but the deep purple flowers in July and August are an absolute picture when seen *en masse*. Indeed hosta flowers are generally much more effective when seen in the mass than as a single spike.

Gertrude Jekyll was a great advocate of hostas as conservatory plants, and she used them just as often in her courtyard schemes. She grew them in Italian terracotta pots to hide what she called the 'common flowerpots' ranged behind. In nearby pots in the shade she grew the male fern and aspidistra and these plants, including the hostas, were often used for filling spaces between the flowering plants, mostly lilies and hydrangeas. She hardly ever composed a courtyard scheme with more than two sorts of plants flowering at the same time and often she just used lilies with her foliage plants. She particularly liked to use hostas because their ample foliage could completely hide the pots beneath.

Terraces

A terrace is basically a 'room without walls' and, as such, is more architectural than horticultural in concept. Plants of strong character will retain the visual balance and, again, those plants which excel in courtyards come to mind – mahonia, bergenia, hydrangea and hostas, but with this difference: in courtyards there is usually some shade, but on terraces usually none, or very little, and the plants must be chosen accordingly.

Although terraces are often sunny and windswept hostas can enhance them enormously and, perhaps surprisingly, some will flourish in such conditions. They look best grown in big, bold groups, planted along the bottom of the house wall or at the foot of the terrace on either side of the steps down to the garden. Hostas associate well with other architecturally strong plants such as yew or thuya which has been clipped or sculpted into square or angular shapes. At the Ladew Topiary Garden in Maryland, USA, *H. sieboldiana* is grown at the foot of terrace steps against the base of a piece of topiary made in thuya. The contrast between the stone steps, the paving, and the roundness and glaucousness of the hosta leaves, against the squareness and bulk of the thuya, makes a simple but highly effective piece of planting (Pl. 26).

There is now a surprising number of hostas which will grow well on terraces, and indeed in America the raisers are deliberately breeding hostas which will withstand full sun. *H. plantaginea* and its hybrids *H.* 'Honeybells' and *H.* 'Royal Standard' are the obvious choices for these positions. Used as border plants, the form of *H. plantaginea* known as var. *grandiflora* has an open habit and very lax leaves, as has *H.* 'Honeybells', and whilst these plants are not in the first rank as border plants, they lend themselves perfectly to terrace planting. Not only do they increase rapidly, filling a long terrace border in a year or two, but their shiny-surfaced leaves reflect the rays of the sun and they flower prolifically. There is also the added bonus that these white or off-white flowers have a fragrance strong enough to waft across the terrace in the evenings. Of the newer cultivars that have been specially bred for their tolerance to full sun, *H.* 'Green Fountain', a hybrid of *H. kikutii*, could be used in tubs on the terrace, as could the even newer *H.* 'Invincible', whose lance-shaped leaves are even thicker, denser and shinier. Some of the gold-leafed hostas actually keep their colour better if grown in the sun and would also be suitable for terrace planting. A pair of specimen *H.* 'Sum & Substance' placed either side of a flight of York stone steps would look spectacular, but would be overwhelming used all the way along the terrace. Of the bright gold-leafed varieties probably *H.* 'Midas Touch', one of the most crinkled and cupped, needs full exposure to the sun in order to bring out its colour.

Tubs with wheeled bases are good for growing hostas in, since they can be moved out of range of the hottest rays of the sun and a much wider variety can then be grown. In common with most other genera of shade-loving plants, hostas flower better when grown in full sun, but at the expense of their leaves, which will be smaller and meaner and will tend to scorch at the edges.

Beds and borders

Beds and borders epitomize the British style of flower and foliage gardening and no one has used them more artistically than Gertrude Jekyll. She made extensive use of hostas in beds and borders, and with apparently only two species at her disposal (and one wonders why she did not use *H. ventricosa*, already grown in Britain for 100 years) she created the most enchanting and original effects. In her book *Colour Schemes for the Flower Garden, Country Life*, George Newnes, 1908, she says of borders '. . . the thing that matters is that, in season, the border shall be kept beautiful, by what means does not matter in the least' and she goes on to list hydrangeas, *Liliums longiflorum, candidum* and *auratum, Campanula pyramidalis* (both white and blue) and, for foliage, *Funkias grandiflora* and *sieboldii* (by which she meant *H. plantaginea* 'Grandiflora' and *H. sieboldiana*), and hardy ferns, as being some of the most useful plants by which to achieve this end.

She used hostas in bold groups, mostly in association with other plants of strong architectural habit, yuccas, bergenias and lyme-grass. Sometimes she used hostas for foliage colour contrast, in front of tree paeonies, for example, harmonizing the glaucous-blue leaves of *H. sieboldiana* with the coppery-purple young leaves of *Paeonia* 'Comtesse de Tudor', with glistening deep green asarum in a nearby corner. Gertrude Jekyll also used hostas in bays between shrubs with the shrubby aster *Olearia gunnii* flowering in mid-June, and with *Lilium longiflorum* flowering in mid-August.

At about the same time as Lawrence Johnstone was experimenting with single colour gardens – and 30 years before Vita Sackville-West's famous White Garden at Sissinghurst Castle – Jekyll was experimenting with and making what she called 'gardens of restricted colouring', by which she meant gardens or long borders in which a particular colour predominates. She used *H. sieboldiana* in her Blue Garden in front of *Delphinium grandiflorum* and *Felicia coelestis* and next to the variegated coltsfoot. In the Green Garden both species of hosta were grown in association with shrubs of bright and deep green, with polished leaf surfaces, green aucuba, skimmia and *Danae racemosus*.

Miss Jekyll knew how to exploit her plant material to the fullest and many of us today would do well to study how carefully, and with what wealth of plant knowledge, she assembled her beds and borders.

Perhaps Christopher Lloyd would not like to be called 'the latter part of the twentieth century's Miss Jekyll', but both great gardeners have a lot in common in the way they use plants. Christopher Lloyd's home, Great Dixter, in East Sussex, is famous for its Long Border which, like the other borders at Great Dixter, is planted with flair and artistry. Lloyd uses stronger colours than Jekyll did, and he uses a wider range of hostas, but he seems, like her, to place his favourites over and over again in different plant associations and groupings in different parts of his garden. *H. ventricosa* 'Variegata' would appear to be his special favourite as on many occasions he has extolled its virtues in print, particularly when it is backed by *Hemerocallis* 'Marion Vaughn', the soft yellow flowers of the day-lily

Fig. 17 *H. helonioides* forma *albopicta*
Description on p. 93

complementing the deep purple flowers of the hosta: and elsewhere *H.v.* 'Variegata' backed by *Veratrum album*, a subtler but equally effective plant combination, or by pink-spiked *Dactylorhiza foliosa*. At the top end of the Long Border *H. sieboldiana* 'Elegans' towers above a froth of *Alchemilla mollis* next to *Rosa pimpinellifolia* 'William III', a striking combination, the alchemilla breaking up the hard line of the stone path.

On the whole, British gardeners are not much given to planting continuous bands of a single sort of plant as a border edging, as is done in the United States, where a number of medium-to-small hostas are used for this purpose, but Christopher Lloyd does sometimes use plants as definite edgings – in short bursts and with great effect, too. But it is his clever use of contrasting background material that takes the stiffness out of edging a border with hostas. He uses elongated fingers of pinky-bronze *Rodgersia pinnata* poking through a long row of *H.* 'Buckshaw Blue', and *Euphorbia palustris* weaving in and out of a sweep of *H. undulata* var. *univittata* to soften what could appear to British tastes a rather regimented line of hostas. Both Gertrude Jekyll and Christopher Lloyd have amply demonstrated that one can achieve stunning effects using only a limited palette of hostas, and without having to seek out the ultra-new and untried varieties.

Hosta walks

Apple Court (The Hosta Walk)

The garden extends to about an acre and a half (0.61 hectares) and is completely walled (it was once the Victorian kitchen garden of the 'big house' next door) and in this we have deliberately set out to provide a special place for hostas – a place where they will not merely do well but will also be a very special feature since, as garden designers, we felt we should not only make the most of our hostas but also make them an integral part of the overall design of the garden. The way we have done this is to devise our own Hosta Walk.

This lies at the foot of a west-facing wall of mellow, rose-red brick. The wall is 184 ft (56 m) long, and the border 9 ft (3 m) deep, with an antique stone bench set on a shallow, rectangular plinth made of the same rose-red brick as the wall (all dug up from the garden – the remains of outbuildings which once existed here) set in herring-bone pattern and surrounded by near-white Purbeck slabs. We tackled the right-hand, more shaded end first, deciding that this was where we should grow mainly hostas with green-and-white variegated leaves, some with glaucous-blue leaves and, as contrast, a few green-leafed species to add crispness and brightness. As we hold the National Hosta Reference Collection of smaller-leafed species and cultivars, these are given prominent positions towards the front of the planting with clumps of larger-leafed cultivars towards the back.

In keeping with the cool green-and-white theme, we are disciplining ourselves to using only a minimum of other genera so that we achieve a restful, rather than a busy, result. We have chosen ferns, astilbes and dicentra to provide a filigree effect, which will contrast well with the substance and solidity of the hostas clumps, grasses which go extremely well with hostas and drifts of phlox cultivars in blue-pink, mauve, purple and magenta. For winter interest we have under-planted with many snowdrop species and hellebores, mainly orientalis hybrids in all shades from off-white to deep plum. Pale silvery-pink roses 'The New Dawn' and 'Souvenir de la Malmaison' will clothe the brick wall, interspersed with mauve-tinted blue-flowered Clematis 'Lady Northcliffe', 'Perl d'Azure' and 'Lord Nevill'. An old damson tree, towards the southern end of this section, seemingly growing out of the foot of the wall, is supporting *Clematis alpina* 'White Moth' and the striped old-fashioned rose 'Variegata de Bologna'.

Five brick pillars holding solid wood crossbeams make a pergola above the hostas casting the necessary half-shade. The pillars are planted with variegated trachelospermum, variegated kadsura, climbing Bourbon roses, *Vitis vinifera* 'Purpurea' and purple-toned viticella clematis. Between each of the pillars is a large terracotta pot containing the good Wisley form of *H. undulata* set in 39 in (1 cm) gravel squares cut into the front brick edging and edged in brick. This repetition gives the whole planting a unified effect and continues with more pillars and gravel squares down the golden border.

Buttresses of yew contain each end of the Hosta Walk and encase the stone bench on its platform, and *Philadelphus* 'Bowles Variegated', *Weigela praecox* 'Variegata' and a group of white martagon lilies add light, airy height and bulk.

This then is the setting for the hostas. We found that many hostas with white margins or central white variegation are classic old favourites like *H. crispula*, *H. fortunei* 'Albo-marginata', *H. undulata* var. *albo-marginata* (whose margins may be slightly too cream early in the season for this scheme), *H.u. medio variegata*, *H.u.* var. *univittata*, *H. sieboldii* and *H. decorata*. Newer ones include *H.* 'Francee', 'See Saw', 'Allan P. McConnell', 'White Christmas', and 'Gloriosa'. The smaller types, *H.* 'Louisa', 'Ginko Craig' and some of the glaucous-blue Tardiana group, are planted along the front of the border and round the base of the pillars.

The north-end Hosta Walk takes on an entirely different colour theme. Gradually the planting becomes more golden-yellow and more spiky and architectural in feeling. With the hostas we have planted phormiums, kniphofias, eremurus, *Crocosmias* 'Citronella', bronzey-grey 'Solfaterre' and 'Canary Bird'. For early spring interest there are tulip species *tarda* and *sylvestris* and the variegated *Fritillaria imperialis*. Clothing the walls are roses 'Gloire de Dijon'; more coppery climbing tea 'Lady Hillingdon' and 'Paul Lede', with buff-yellow 'Alchemyst' and 'Phyllis Bide' on the pillars. Greeny-white double-flowered clematis 'Duchess of Edinburgh' echoes the slate-grey double clematis 'Belle of Woking' in the silver border. Yellow, orange and tawny day-lilies and yellow regale lilies add flower power from

May onwards and to unify both sections of the walk, *Phlox* 'Prince of Orange' and 'Brigadeer' will emphasize the phlox theme.

The dazzling array of hostas with cream/yellow/gold-margined leaves and central stripes will come into their own here. The magnificent *H. montana* 'Aureo marginata' and 'Spinners', 'Antioch', *H. fluctuans* 'Variegated' as well as 'Phyllis Campbell' and 'Elizabeth Campbell', interspersed with the larger gold-leafed types, 'Golden Sunburst', 'Semperaurea', 'Gold Seer', 'Zounds' and, most sumptuous of all, 'Sum and Substance', with thick, waxy leaves as an endstop, grows in full sun, even withstanding the reflected heat from the brick wall. Large lance-leafed, lime-green 'Green Acres' and spinach-green *H. ventricosa* add sharpness and spice to the background. The sections of the border between the pillars and the walls provide shade for about half the day and are perfect for those gold-leafed hostas which cannot withstand hot sun. Here we are planting 'Sun Power' and 'Piedmont Gold'. 'Gold Standard', 'Janet' and 'Moonlight', whose changing coloured leaves will also benefit from this shade, are planted behind the pillars too. 'Shade Fanfare', 'Nancy Lindsay' and 'Golden Tiara' will complete the middleground, while for the front row, out of a huge possible selection, we have chosen little green-leafed *H. kikutii*'s var. *polyneuron* and var. *caput-avis*, *H. helonioides* 'Albo-picta' and the gold-leafed edging hostas 'Gold Edger', 'Golden Prayers' and 'Midwest Gold', a dull gold and one of the earliest hybrids ever raised with gold leaves. The dwarf hemerocallis, yellow 'Eenie Weenie' fits in well here with the grass *Hackonechloa macra* 'Albo-aurea' which is such a good foil for hostas.

Our hosta walk is a new planting and there will inevitably be some fine tuning of the plant arrangements for some years. The 'walk' itself is an ancient concept in garden design. Hosta walks are not new either and we must admit to being inspired by two enchanting examples, each totally different but both ideally suited to the growing of hostas.

Hadspen House

The famous Hosta Walk at Hadspen House in Somerset has been created where Eric Smith originally had his hosta seedling beds and later grew his Tardianas and gold-leafed hybrids. It draws hosta enthusiasts from all over the world.

The grounds of this hamstone manor house are situated in a south-facing bowl surrounded by hills, with tall trees to the north, and the hosta walk is located in the old, partly walled, vegetable garden which is sheltered from extremes of climate.

The soil is heavy alkaline clay and very moisture-retentive, which accounts for the fact that the hostas grow quite successfully in full sun. However, in order to provide some degree of shade, the beech hedges backing each side of the hosta walk will be trained up and pleached. Purple-leafed pittosporum and carmine-coloured rugosa roses grow in front of the beech and provide a pleasing colour contrast to the many glaucous-leafed hostas. The slab path down the centre is slowly being encroached on by the hosta clumps as they increase in girth. The white-painted wooden Lutyens seat at the bottom of the walk draws one's eye down the complete length and is a perfect foil for the hostas. The general impression given by the hosta walk is that of a sea of blue and yellowy-gold.

The Walled Garden

The Hosta Walk at The Walled Garden, the historic garden which was the Vanbrugh-walled vegetable garden for Claremont Park in Surrey, runs the length of the shaded section of the 15 ft (5 m) high wall. Deep rose-pink brick is enhanced by a curtain of slightly paler pink *Clematis montana* 'Rubens', deep pink Japanese azaleas and a planting of large clumps of *H. sieboldiana* 'Elegans' fronted by repeated clumps of *H. undulata*. An air of coolness and continuity prevails, but the cerise blooms of the nearby Judas tree add completeness to this simple but effective scheme, which shows what can be achieved if hostas are used with such restraint.

Greyrigg – hosta growing Japanese style

Dr Dilys Davies's garden on the eastern fells in the Lake District in Cumbria, is situated on terrain that many of us would consider ungardenable. Three-quarters of her garden is at an angle of 45 degrees or steeper and consists of rock with a shallow layer of soil; the main reason for growing hostas on this cliff is to stop the earth falling off. Quite by chance Dr Davies has stumbled on a way of using hostas on an English hillside that remarkably resembles the way they grow in the wild.

She has created, against these odds, a remarkable garden by any standards, probably quite unique in Europe.

Greyrigg is situated 600 ft (183 m) up the side of a mountain composed of degraded granite over which lies a thin, peaty and extremely acid soil, varying in depth from 2 in (5 cm) to $2\frac{1}{2}$ ft (75 cm). The garden is made up of vertical and near-vertical rock faces interrupted by occasional shelves. A small stream runs through the garden. The garden faces due west, though wind is not a particular problem. The average rainfall is about 120 in (30 cm). However the soil dries out quickly, partly because the steepness of the site naturally drains well, but also because the underlying rock heats up in hot weather and holds the heat, drying not only the earth but also any plant roots in it – acting like under-floor central heating. The hostas do not seem to mind this. There are pockets of boggy soil, and places where the earth is almost always dry. The only really deep soil is the bog at the bottom of the garden.

The Davies's started using hostas right from the beginning. The first was *H. sieboldiana* which Dr Davies originally thought was var. *elegans*. She likes its very large, greyish leaves and the flowers which rise above the leaf-mound and make attractive dead heads in winter. She has also used this hosta to line a bed in which a bamboo and a 4–5 ft (1.4–1.5 m) kniphofia are grown. The front of the bed is edged with *H. undulata* var. *medio variegata* (nomen nudum), this being her name for the particular plant she grows. The original plants of *H. sieboldiana* have now gone on to what Dr Davies calls the 'ski slopes' – very narrow, shallow ledges on the cliff face. They have had three seasons up there and are now quite luxuriant, even suppressing *Lamium galeobdolen* 'Variegatum'. She also used *H.* 'Thomas Hogg' to begin with and this is now dotted round the garden.

Since those early days in 1971, other hostas have been acquired, many of them large varieties. One section of the cliff is covered with soil. It is fairly steep without any retaining ledges so she has not been able to terrace it. This area has been planted with some very big clumps of *H. sieboldiana* and *H. fortunei* 'Albo-picta' and even the resident mole has been unable to dislodge them.

One remarkable piece of hosta planting came about almost by accident. A piece of one of the *H. fortunei* family was left over from another planting and Dr Davies just left it on a very shallow piece of rock meaning to build a bed round it. It was a decent-sized clump which stayed in place of its own volition, and she simply forgot it. It has been there for anything between 10 and 15 years sitting entirely in its own roots, growing bigger and better each year and flowering exceedingly well.

Perhaps partly as a result of that happy chance, Dr Davies deliberately started planting in 1985 what is now known as the Hosta Cliff. She planted *H. lancifolia*, an *H. longipes* raised from seed, *H.* 'Thomas Hogg', *H.* 'Royal Standard', an unnamed *Tardiana* seedling, and a variegated *H. fortunei*.

Dr Davies had many of the Tardiana group from Eric Smith in the early days and all have flourished in and suit their surroundings except for *H.* 'Blue Moon', known then as *H.* 'Halo'. *H.* 'Blue Diamond' obviously likes the soil and the high rainfall and, like all the Tardianas growing at Greyrigg, is bluer than Dr Davies has seen them elsewhere. She puts this intensity of colouring down to the continuous rain.

All the hostas do exceedingly well. The only ones she has lost, apart from *H.* 'Blue Moon' which was eaten by slugs, were the micropropped *H.* 'Reversed' and *H.* 'Hoopla'. *H.* 'Borwick Beauty', the reverse variegation of *H.* 'Frances Williams', although initially slow, is now clumping up very well. *H.* 'Buckshaw Blue' is planted on some of the ledges in the shade where the leaves are not scorched and again they are doing their job of holding soil on the hillside.

Along the stream bank there is a planting of *H. tokudama*, along with a red acer, some rodgersias and ligularia, and beyond these a large chunk of *Iris pseudocorus* 'Variegata'. It all looks beautifully luxuriant and the water trickling past is an added delight.

Vicar's Mead – a hillside garden

It is, perhaps, almost typical of a plantsman's garden that the terrain should actually be difficult to garden. This is certainly the case at Greyrigg. Vicar's Mead is, again, a garden where much of the interest arises from the skilful use of good plant material to overcome not only topographical difficulties but also climatic difficulties, for the garden, though situated in an area of generally high rainfall, actually receives very little rain.

Vicar's Mead, a 500-year-old Devon 'long house', notable as the house in which Sir Walter Raleigh received his education, is situated at East Budleigh, on a sandy, heathland peninsula jutting into Lyme Bay in the English Channel, near Exmouth. The garden consists mainly of a north-facing cliff, some 100 ft (30 m) above sea level, linking two areas of flat ground, a large, roughly square area of about one acre (0.4 hectares) containing the house, and a narrow, fairly level strip at the top of the cliff.

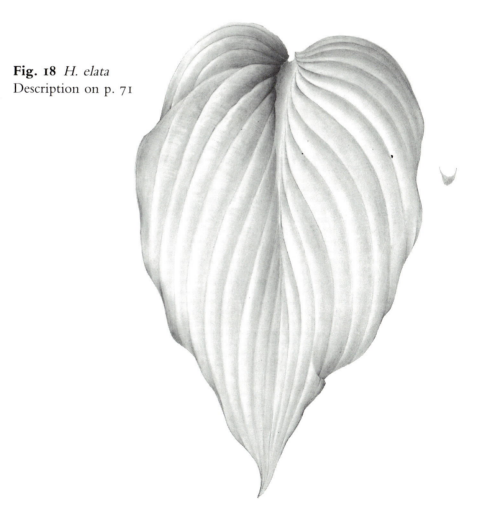

Fig. 18 *H. elata*
Description on p. 71

This red standstone escarpment creates a curious physical environment in which to garden, for, on the one hand, it shelters the garden from the prevailing westerly winds, but on the other, by sheltering the garden, holds off the rain. Another two-and-a-half acres (1 hectare) of cultivated terraces and slopes rise up above the house.

When Liz and Harold Read moved here in 1977 they had to re-design and re-plant the garden from scratch in order to accommodate their large and varied plant collections. The soil is a just-acid (pH 7) sandy loam. It has to be mulched heavily with compost, manure and forest bark to retain the moisture.

In spite of these difficulties, hostas grow here luxuriantly and are one of the special features of this very popular and well-known garden. The extensive hosta collection is planted on a small plateau just above the systematic beds containing the National Reference Collections of liriopes, ophiopogons and libertias. The hosta planting faces south-east and there is only shade during the hottest part of the day, and even less now that the big ash tree, which formerly provided shade, was struck by lightning.

The hosta bed is backed by trees and tall shrubs, in front of which are herbaceous plants chosen to complement the hostas – cimicifuga, dicentra, astrantia, *Anemone japonica* and polygonatum and, on the sunny side, phormium, astelia, libertia and Pacific Coast hybrid irises. The hosta bed is formed in the shape of an elongated horseshoe with a narrow, Devon slate path creating the outline. The backbone of the planting is a huge clump of *H. sieboldiana* 'Elegans', flanked at each end by *H.* 'Frances Williams' and *H. tokudama* 'Flavo circinalis', with *H.* 'Sum and Substance' at the open end. Between the path and the central clumps are *H.* 'Louisa', 'The Twister', 'Golden Sunburst', some Tardiana hybrids and 'Pastures New', *H. longissima* and *H. fortunei* 'Hyacinthina Variegata'. At the outer edge of the path other small hostas are grown, including 'Lemon Lime', 'Shade Fanfare', a group of very small unnamed seedlings, and an *H. venusta* type, 'Groundmaster', 'Ginko Craig', 'White Christmas', 'Kabitan' and *H. rupifraga*. The outer ring contains large specimens with *H.* 'Krossa Regal' and *H.* 'Snowden' as a matching pair of greyish-leafed specimens planted centrally opposite each other.

The whole planting has been carefully balanced by Liz Read, who likes to blend leaf-shape, colour, texture and size with variegation and solid colour. Liz has been involved for many years with the National Association of Flower Arrangement Societies so chooses her hostas primarily as flower-arrangement material. None the less, as Vicar's Mead is open to the public under the National Garden Scheme, and for other charities and horticultural societies, every plant in the garden must earn its keep.

The hostas have adapted well to the quick-draining, sandy soil and, apart from *H.* 'Sea Drift' which became virused, the hostas are all healthy specimens which increase well. Slugs are a problem (as with most Devon gardens), so slug pellets are regularly applied, as well as intermittent doses of 'Nobble'.

Spinners

Spinners, at Boldre, Hampshire is a small woodland garden, where hostas play an important part in the planting. Owned and gardened since 1960 by Peter Chappell, a former schoolmaster, and his artist wife, Diana, Spinners now incorporates a well-known specialist nursery – a mecca for keen gardeners.

The garden has been created beneath the high shady canopy of oak trees which provide the ideal habitat for hostas and other shade-loving plants. The upper part of the garden is on New Forest gravel, a poor soil which has to be supplemented with heavy annual mulches of fallen leaves, but it suits the species rhododendrons, azaleas, camellias and hydrangeas among which the hostas are grown. The lower reaches of the garden are on sticky, yellow clay.

The entrance to Spinners is down a long sloping drive along the left-hand side of which a variety of hosta species and cultivars are planted, benefiting from the dappled shade of the oaks to the south side. *Acer palmatum* 'Atropurpureum', bluish-pink-toned rhododendrons and a magnificent specimen of *Rh. augustinii* 'Electra' greet the visitor, whose eye is soon caught by a large clump of *H. undulata* 'Erromena', and then a mature 'Birchwood Parky's Gold' which becomes increasingly golden until September; its flower spikes are already well-advanced by mid-May. Nearby are several interesting seedlings given to Peter by Joy Bulford who knew Eric Smith when he was working at Hilliers of Winchester. Her *H. kikutii* looks true to type and is a very classy small hosta. The edges of this border are flanked with long drifts of *Tellima grandiflora*, *Epimedium perralderanum* and *E. versicolor* 'Rubrum', behind which *H.* 'Krossa Regal' towers next to groups of choice ferns. Among these is a hosta given to Peter Chappell in the early 1970s by Eric Smith, who stressed that it is *H. sieboldiana* var. *aureomarginata* – not *H.* 'Frances Williams', which it generally resembles. Peter divided the plant a few years later and, for comparison, planted it next to what is generally and almost universally sold as *H.* 'Frances Williams', a slightly less attractive plant with narrower, less distinct margins and rounder leaves. That these two are quite distinct cannot be disputed when seen growing next to one another, but the differences are well within the tolerances for the species. The problems arise because there is an ICBN rule which states that where similar sports arise from the same species those similar sports should all bear the same name regardless of origin. Thus all gold-edged sports of *H. sieboldiana* are 'Frances Williams', just as the solid gold sports from 'Frances Williams' are 'Golden Sunburst' (p. 112). However, some forms of 'Frances Williams' have far wider gold margins than others and are definitely superior. No doubt the Eric Smith plant was one of these, as was the form once known as 'Golden Circles', possibly they were the same.

Moving down the long drive border towards the house, the hosta planting becomes more dominant with massive clumps of *H.* 'Snowden'; *H.* 'Hadspen

Blue' grows with *Tiarella cordifolia*, a happy plant association; Tardiana *H.* 'Peter Pan' next to *H.* 'County Park' seems to thrive here in the dryish, woody soil. Glaucous *H.* 'Pastures New' looks well with shiny hart's tongue ferns and *Adiantum pedatum*, with *Saxifraga fortunei* and *S. stolonifera* making good foils for another patch of Tardiana hybrids sent to Peter many years ago by Netta Statham. Japanese anemones for autumn colour and plum-toned *Helleborus orientalis* for winter add cheer to the textured tapestry of foliage and subtle flower colour. Early spring is heralded by small bulbs whose foliage has died down before the hostas come into leaf.

Returning half-way up the drive one turns left up a narrow, grassy glade with planting of *Pieris* × 'Wakehurst' and a huge *Rhododendron williamsii* hybrid, to where flourishing clumps of *H.* 'Hadspen Heron' and *H.* 'Chelsea Babe' make a delightful combination near the best form of *Rh.* 'Bow Bells', cerise in bud and pale pink in full flower. On past *Rh. loderi* 'King George' and some Pacific coast hybrid iris, one comes across diminutive *H.* 'Hydon Sunset' running everywhere, with slow-growing *H.* 'Gold Haze' behind, and equally slow-growing *H.* 'Janet' across the path. *H.* 'Weihenstephan', brought to Britain by Dr Ullrich Fischer, is much more vigorous here and is beginning to show signs of rugosity in the leaves after only three years. The corner is turned to face yet more – and each one distinct – Tardiana seedlings in what Peter says is his best foliage border; a sea of sunshine-yellow in May with trollius, *Filipendula ulmaria* 'Aurea', *Lysimachia nummularia* 'Aurea' and drifts of yellow spurge, *Rheum palmatum* 'Red Herald' towering above them. Across the grass the yellow theme is echoed with new plantings of *H.* 'Zounds' and *H.* 'Golden Sunburst' lighting up a dark corner. Here *H. plantaginea* was only just poking through the ground (Spinners is in a frost pocket), and even though higher up the garden, its hybrid *H.* 'Honeybells' was already in full leaf. In this summer-foliage area is a good-sized planting of *H.* 'Buckshaw Blue'.

H. 'Blue Angel' was showing its full potential after only three years: huge, shapely, heavily pruinose, greyish leaves on thick, wide, fleshy petioles. It was proving to be one of the best of Paul Aden's many named *H. sieboldiana* forms: far better and more interesting than *H.* 'Big Daddy'.

Back to the small pond near the sales area, fed by a spring in the garden. In one corner a magnificent cupped and puckered, deep glaucous-blue *H. sieboldiana* form (it must have *H. tokudama* in it) is backed by *Matteuccia struthiopteris*, a huge hybrid Lysichitum, two different forms of variegated water iris and pale-pink-flowered *Rh.* 'Daydream'. This pond area is at its peak when the ice-cold, clear blue meconopsis come into flower in late May.

On the downhill slope, the soil is thick, heavy clay with an herbaceous border to the left. It is at its peak in July when agapanthus and crocosmias mingle with *Rosa* 'Ballerina', into which mature clumps of *H.* 'Spinners' and *H. crispula* nestle. Both are susceptible to wind-bruising early in the year, *H. crispula* the more so. Since it comes though the ground before most other hostas, it is also often damaged by late frosts, and its curling, twisted leaf tips can be ruined for the whole season.

Moving along this border further into the shade, one comes to a sheltered corner between huge *Rh. ponticum* and forest oaks. On ground sloping south, and open to sun from the west in the late afternoon, there is a planting of a whole batch of Tardiana hybrids which Eric Smith deposited here for safe-keeping and never came back for.

The Savill Garden, Windsor

The gardens at Windsor Great Park epitomize the art of the English woodland garden and they provide one of the best settings for hostas, especially the part now known as the Savill Garden – where hostas are grown mainly in large drifts, or as bold plantings of ground cover.

The gardens are situated on the south-eastern boundary of Windsor Great Park which itself extends to some 4500 acres (1821 hectares) and is a remnant of the once vast, primaeval Windsor Forest. The soil is acid, sharply-draining Bagshot sand, and the main planting throughout is of acid-loving genera, with rhododendrons and azaleas predominating. The climate is not severe, though the rainfall is low, only 22 in (55 cm). The ground is sharply contoured, with hills and valleys interspersed by streams and ponds. The gardens lie in dappled shade cast by a high canopy of huge, ancient oaks and groves of grey-stemmed beeches. The plants thrive in spite of the low rainfall because, from the very start, effective irrigation was installed.

The gardens as we known them today were largely the creation of Sir Eric Savill who was appointed Deputy Surveyor of Windsor Great Park in 1931 and later became Deputy Ranger. Originally the Savill Garden was known as the Bog Garden, and one could enter it, without charge, by climbing a stile.

Here, hostas, mostly the larger-leafed types, play a secondary but useful role, filling large spaces between the shrub masses. *H.* 'Anticipation', a seedling which arose here is much used. It is presumed to be a hybrid between *H. sieboldiana* and a hosta of the *H. fortunei* complex. Although it has many of the best characteristics of both parents – glaucous, mid-green, cupped and puckered leaves – the Keeper of the Gardens, John Bond, says, 'It is a useful ground-cover plant but not a world-shattering hosta', and it is used mainly because it arose in the gardens. Likewise, *H.* 'Journeyman', similar in size and leaf colour to *H. undulata* 'Erromena', but with rather prettier flowers on a 2 ft (60 cm) scape. But the garden has always had a special interest in hostas, along with hardy ferns, as ground-cover contrast plants, and hostas are used not so much for their own sake but more for the benefit and enhancement of other plantings. Indeed it was Sir Eric Savill who named *H.* 'Tallboy', that excellent, tall-scaped hybrid of *H. rectifolia* (now sadly neglected), long before it was fashionable to take an interest in hostas.

John Bond VMH has been the Keeper of the Gardens at Windsor Great Park since 1970. He has considerably increased the range of hostas grown there. When he took over the only hostas there were *H. sieboldiana*, the *H. undulata*s, *H. lancifolia* and the *H. fortunei*s. Over the years he has been given hostas from well-known plantsmen and women, including Nancy Lindsay, Gwendolen Anley and Eric Smith. He now also acquires them commercially in order to keep up to date with the latest sorts. He uses them in creative and imaginative ways and on a scale which would be quite impossible to emulate in most private gardens. There is no doubt that they lend themselves perfectly to this sort of landscape and they are seen at their best when used on this scale. The plantings at Windsor are similar to those on Battleston Hill at Wisley some 20 miles (32 kilometres) away, from where many of them came. Interestingly, they fend well for themselves used like this, and even in the great drought of 1976 none of the established hosta plantings flagged at all, even in the driest, sandiest part of the gardens.

All new plantings are given farmyard manure and there is no doubt that the hostas benefit from this. Every year, in common with all the plants in the woodland areas, the hostas are mulched with a deep layer of rotted leaves, which is supplemented by the natural leaf fall. This method of cultivation seems to suit the hostas very well indeed. *H. decorata*, a stoloniferous hosta, and notoriously difficult to grow successfully in Britain, thrives here in the peat bed.

Hostas have for a long time been associated with particular areas of the gardens and were mentioned by Lanning Roper in his book *The Gardens in the Royal Park at Windsor* (Chatto & Windus, 1959), several years before hostas became at all well-known to British gardening. Lanning Roper refers to hostas being used as underplantings to the species collection of rhododendrons in conjunction with the royal fern, epimediums, mahonia, Solomon's seal, violets and ericas (an unusual combination!). He says that hostas, with their distinctively shaped leaves, create a contrast to the heavy foliage of the rhododendrons.

Today the most successful planting of hostas at the Savill Garden occurs where they are used in conjunction with huge drifts of the Himalayan blue poppy, *Meconopsis grandis*, its blueness accentuated by juxtaposition with the orange *Primula chunglenta*. Beside the Himalayan blue poppies, in an open glade beneath towering oaks and a fringe of deciduous azaleas, are bold plantings of *H. sieboldiana* and *H. fortunei* 'Aurea', their substance and solidity a foil to the airy, delicate sky-blue poppy flowers (Pl. 37). At the opposite side of the glade this theme is echoed by a massed grouping of glossy green and gold-leafed *H. ventricosa* 'Aureo maculata', backed by more meconopsis and *Matteuccia struthiopteris* (the shuttlecock fern), with a huge planting of the late Mrs Gwendolen Anley's soft golden-leafed *H. sieboldiana* 'St Georges Gold'. This particular form of the golden-leafed *H. sieboldiana* was, according to John Bond, growing at the Savill Garden long before golden-leafed hostas became popular or widely grown. Although this planting, using only five or six different hostas with the Meconopsis, is an ephemeral rather than a seasonal splendour, it is a magical horticultural experience and surely one of the most outstanding at the Savill Garden – successful enough in the opinion of John Bond to go to the trouble of re-planting the meconopsis every third year.

An American garden created in Japanese style

Hostas are equally popular on both sides of the Atlantic, but the ways in which they are grown differ substantially for reasons which are partly climatic and partly cultural. It is the intensity of the sunlight that more than anything else dictates how hostas are grown in America, for there it is so strong that hostas have to be grown in shade: yet the shade is often so deep that one would scarcely expect them to survive at all. The saving grace is the high summer air temperature which, when they are given enough moisture, causes them to grow much bigger and faster than they do in British gardens. These extremes between light and shade limit the range of other plants which can be grown with hostas to provide a foil for them. This is one reason why many American growers and collectors seem happy to give over their entire gardens to hostas.

The most remarkable hosta garden known to the author personally in America is that of Warren and Ali Pollock. Dr Pollock is a former editor of the *AHS Bulletin*, Chairperson of the AHS's Awards & Honors Committee and also former Chairperson of their Nomenclature Committee; a keen hosta grower and collector of many years. The Pollocks's garden is in Wilmington, Delaware, on one-third of an acre (0.1 hectares). Their garden won the 1987 Design Plaque Award of the American Hosta Society for the best landscape use of hostas in a home garden.

Seen through the eyes of one used to British gardens, one of the most extraordinary things about the Pollocks's garden is that hostas are just about the only plant material used – apart from grass and the trees that were already there. The trees are mainly native red maples with greedy root systems, and Warren explains that when he had the house built about 20 years ago he decided to keep as many trees as possible, with just a narrow, tree-free strip round the house. As a result, most of the garden is in dense shade all day long while at the edges the hostas hover, as if illuminated by an irridescence of their own. In fact, in order to grow hostas at all, the Pollocks have to undertake extensive site preparation, digging huge individual planting holes for every hosta, sometimes removing quite large tree roots in the process, and then enriching the return soil. They then have to water and feed the hostas regularly.

The Pollocks have used some 200 hostas, many in large masses and drifts of the same species or cultivar, to create scores of different landscaping themes. They have utilized some of the larger trees as 'dividers' to develop 'rooms', each area planted with special hostas for particular effects of colour and texture. To set off some of the designs and to separate them from adjacent plantings, many evergreen shrubs, mostly green-leafed euonymus, were planted, and they provide the winter scenery.

Top soil was brought in and piled into mounds, each strategically placed and decorated with hostas to provide a special view (Dr Pollock uses the word 'picture') from a house window, a walkway, the rear porch or the raised terrace in the back garden.

This arrangement creates not only a perfect display setting for the hostas but also a remarkable garden. Nothing in the design was left to chance: it is a thoughtful, even cerebral garden, and perhaps it is to this that it owes its harmony and serenity.

Dr Pollock adopted at the outset the philosophy of aesthetics to which he has adhered throughout the making of the garden. He has used several ancient, fundamental Japanese landscaping principles to make his garden appear bigger than it really is. Gardens in Japan, especially city parks and the landscaping around many shrines, are designed so that one rarely sees more than one garden picture at a time. The paths are not straight and one can never see round a corner. At almost every step along a path, the eye sees something different. As a result, the pieces comprising the landscape picture seem to add up to an illusion greater than the sum of what actually is there. Dr Pollock also avoids planting along the boundary as that would stop the eye. By carefully positioning accent plants and a few garden ornaments, he leads the eye away from the end of his property. Paths seem to go nowhere. He can borrow views of his neighbours' gardens because the boundaries are not defined, creating the illusion that they are part of his own. Two paths seem to wander into his neighbour's garden which is also heavily wooded. He uses different textures and shades of green to create the illusion of depth, as do Japanese gardeners and artists who paint landscapes. Big, bold, dark green, heavily textured hostas like *H. nigrescens* are planted in the front with small, light green ones in the background, which makes the distance between the two seem greater than it actually is.

In the front garden there is an island of trees planted with hostas. The planting has been cleverly done so that one cannot see it at all from the road and it is actually appreciated best from the house looking out.

As one comes into the garden from the road and then goes up to the front door, four different 'outside rooms' are revealed, one after another. Each setting comprises a different arrangement of hostas, designed to complement each other and to take advantage of the various tree locations. First, one sees a mound planted with two big, colourful clumps of *H.* 'Antioch' with a *H. fortunei* 'Aureo-marginata' clump between them in the rear and a small yellow-leafed hosta in the centre front (Pl. 36). One then sees a large triangular-shaped mound, the 'Yellow Bed', planted with two dozen yellow and gold-leafed hostas, as well as with gold-leafed white-edged varieties including *H.* 'Anne Arrett', 'Lunar Eclipse', 'Moonglow', and 'Moonlight'. These are mostly small or medium-sized, with the smallest at the front. Just as the White Garden at Sissinghurst Castle is not entirely a white garden, so this bed is not all yellow or gold. The chartreuse-leafed *H.* 'Sum and Substance', a marvellous, giant-sized cultivar, is used to tone down the brilliant yellows and golds and to stop the eye. Two clumps of *H.* 'Allan P. McConnell', with small, long lance-shaped, white-bordered, light green leaves, are on either side of the mound near the front (Pl. 36).

Further on up the path there is a group of big, bold, blue *H. sieboldiana* 'Elegans', and between them, the smaller, gold-leafed *H.* 'August Moon', slightly to the rear. 'August Moon' is planted on a mount to give balance to the design. *H. sieboldiana* 'Elegans' and *H.* 'August Moon' have similar leaf texture and shape, and are good hostas to use together.

Finally, one sees an arrangement of five big clumps of *H.* 'Frances Williams' in a flaring half-circle, with a specimen-sized gold-leafed *H.* 'Piedmont Gold' in the centre at the front of the bed. These two hostas have markedly different leaf shapes and textures, and the contrast works well.

To complete the picture, at each side of the front porch are two clumps of *H.* 'Hadspen Blue'. These two glaucous-blue-leafed small hostas, used in this way, as a pair, act as what landscape architects would call 'markers' and announce that you have reached your destination and may ring the door bell. For a garden like this is not merely there to be looked at; it is an experience in which ones takes part as one passes through it.

Longstock Water Garden

Hostas could perhaps be regarded as bog plants, for while most of us cultivate them in ordinary garden soil, there is no doubt that they thrive at the water's edge. Not only do they grow into exceedingly large and luxuriant plants if they have their roots in moist soil, but somehow they look at home growing in a ribbon along the banks of a stream or at the side of a pool. At Longstock Water Garden this potential is revealed to the full.

Longstock Water Garden, Hampshire, was the brainchild of John Lewis, the founder of the department-store empire, who gave his mansion and grounds to his retired employees as a holiday home. The 7-acre (2.8 hectares) garden lies in a wide, sheltered valley on a seam of greensand between the chalk of Andover and Stockbridge, on a tributary of the famous trout-fishing river Test. It consists of a shallow lake which gives way to a series of small, interconnecting streams and pools which were once watercress beds, on the banks of which are grown a wide variety of waterside plants.

Here there is a selection of plants of architecturally strong habit, including grasses and sedges which associate particularly well with the hostas, as do the many hardy ferns grown here. However, the primulas make the greatest impact, and they are at their peak in June, even though they compete with astilbes, rodgersias and day-lilies for intensity of colour. Cerise-pink is the predominant colour, echosed by the dark red leaves of the maples. Later in the year the *Lobelia cardinalis* hybrids continue this colour theme. A broad riband of *H. sieboldiana* (at least 30 ft (9 m) long), planted along the edge of the largest of these streams, makes a wonderful foil for the varying shades of red and pink (Pl. 35).

The hostas used here are those that have achieved a high measure of popularity because they are good garden plants and they are used in broad sweeps along the banks or in bold deep blocks behind the waterside plantings. An especially striking form of *H. fortunei* 'Albo-marginata' grows in several different parts of the garden. In some areas, mainly in deep shade, its leaves are very broadly streaked with white, while elsewhere only the margins are white and very narrow. *H. undulata* 'Erromena', not now considered a top-ranking hosta, looks surprisingly good when grown *en masse*, for the leaves have a clearly defined shape, and the clarity of their shape is reinforced by repetition and by reflection when grown at the water's edge, an effect all the more intriguing when the hostas are in flower, their tall scapes leaning out across the water, curving up and away from their inverted images. (Pl. 34) *H. sieboldii* is equally effective used in this way and both benefit from the quality of the accompanying planting.

The garden has been designed to be seen from all sides and from the vantage point of a small summerhouse near the middle which is reached by a series of wooden planks over the streams. The possibilities of plant association are therefore endless, and hostas, as the garden is seen from different points of view, are variously the feature and the foil.

National Hosta Reference Collections

In common with the majority of ornamentally important genera growing in Britain, there are National Reference Collections of hostas. These collections are over and above those held at the Royal Botanic Gardens, Kew, and other botanic gardens (which are limited mainly to species) and are run under the auspices of the National Council for the Conservation of Plants and Gardens.

National Collections are comprehensive collections of specific plant groups, brought together in designated gardens, public, private and institutional, to serve as living reference points for identification, comparison, propagation, distribution and further breeding and research. They are the horticultural equivalent of libraries and museums. The National Hosta

Collections are held by the Royal Horticultural Society in their garden at Wisley, Woking, Surrey; by Leeds Parks Department, Leeds, Yorkshire (large-leafed species and cultivars); at Apple Court, Lymington, Hampshire (smaller-leafed species and cultivars) and at Kittoch Mill, Carmunnock, Glasgow. Some of these collections and their locations are described in this chapter. All these collections are at various stages of development, but are still very young since the NCCPG did not begin its National Reference Collection scheme until 1980. All collection holders co-operate with each other over the acquisition of plants and in other matters relating to the cultivation and management of their collections and any problems involved. As far as hostas are concerned, probably more than most other genera, the hope and expectation is that the problems of identification which have arisen since hostas were first grown as garden plants, will be resolved. For each species in the collection there will be a standard herbarium sheet and there will also be an equivalent sheet for each cultivar, known as a standard specimen. These sheets will be held at the Royal Horticultural Society's laboratory at Wisley and will then always be available for consultation by interested parties – nurserymen, for example, wanting to make sure of the identity of their stock plants. In due course, in a perfect world, no hosta

Fig. 19 *H. kikutii* var. *polyneuron*
Description on p. 81

should change hands incorrectly named and labelled. We still have a long way to go but at least a very good start has been made.

The National Hosta Reference Collections are located in different parts of Great Britain geographically and climatically and their sites are, on the whole, dissimilar. The principle and largest collection is held by the Royal Horticultural Society at their garden at Wisley in Surrey.

The hostas were originally grown on Battleston Hill, a high ridge of ground, wooded with Scots pine, oak, sweet chestnut and birch. The northern slopes were cleared in 1937 and planted with rhododendrons, around which large clumps of *H. sieboldiana* and *H.s.* 'Elegans' grew and flourished, in company with *H. sieboldii*, *H. plantaginea* and several hostas of the *H. fortunei* complex. Camellias, hydrangeas and day-lilies grew in company with the hostas, as did *martagon*, *regale*, *tigrinum* and *henryi* lilies, all revelling in their protective high oak canopy, letting in only filtered sunlight.

Over the next four decades hostas grown from seed sent from Japan (often received incorrectly named), were planted with the newer cultivars as they were raised and sent to Wisley from American nurseries and hosta collectors. *H.* 'Honeybells' and *H.* 'Royal Standard' and later *H.* 'Frances Williams' created great interest from the gardening public when they were first planted. The comprehensive collection of hostas grown at Wisley was a particular interest of the then Director of the garden, C.D. Brickell VMH, now Director General of the Royal Horticultural Society, who is President of the British Hosta and Hemerocallis Society. It was he who was responsible for the RHS's exhibit of hostas in the Scientific Section of the Chelsea Flower Show in 1968, which was really the departure point for current hosta popularity in Britain (Pl. 30), and it was he who originated the concept of National Collections.

The majority of hostas in the collection are now grown in the Wild Garden, the oldest part of Wisley garden, an area of cultivated woodland between the Heather Garden and the Rock Garden, in a flat, shallow valley where the main plantings are of species and cultivars of rhododendron and other ericaceous shrubs and trees and where there is a high canopy of mature oak trees. Many of the large trees were blown down during the storm of October 1987 and there are now more open glades than there were formerly. The soil is Bagshot sand, which is very acid and fairly poor in nutriment, so that the hostas do not exhibit the vigour and luxuriance one might expect were they grown on heavy clay. Also this is a part of the garden designated as 'wild' and the plants, up to a point, have to fend for themselves. On the other hand there are at Wisley facilities and resources available to ensure accurate record keeping and it is here that the first standard and herbarium specimens (p. 58) will be housed. The greatest problem at Wisley is the theft of plants. One or two particular cultivars are stolen so frequently that many visitors never get the chance of seeing them as mature plants. In common with all National Collections, there must always be a supply of back-up plants available in case one dies or disappears. There is also a hosta propagation area (not available for inspection by the public) where stocks of newly acquired plants can be built up for eventual distribution to the nursery trade and other collections and gardens.

The number of species and cultivars grown at Wisley is already quite comprehensive and is always increasing since over the years botanic gardens, nurseries and collectors from Britain and other countries have donated hostas. Most of them come from Japan and America. It is also here at Wisley that most of Eric Smith's Tardiana group is growing – in fact Smith personally donated a piece of each plant in his collection – and it is here at Wisley that the British Hosta and Hemerocallis Society's Nomenclature Committee, headed by Chris Brickell, is starting to identify and name those Tardianas still only known by number. There are still hostas in the National Collection designated as *Hosta* 'species' because they have not yet been identified, but no doubt they will be assessed in time.

Apple Court

In addition to our general interest in growing hostas of all sorts, the author has a special responsibility for the smaller species and cultivars since she holds this section of one of the National Reference Collections.

Apple Court lies on the south coast of Hampshire. On the whole the climate is equable with an average rainfall of 32 in (105 cm), and plenty of sunshine as the area is protected from much of the summer rain by the Isle of Wight. The soil is a well-worked medium loam, overlying heavy, sticky clay which is perfect for the cultivation of most hostas. The garden was once a chrysanthemum glasshouse nursery and later grew strawberries commercially, the fruits being sent up to Buckingham Palace.

While the main hosta planting is concentrated in the Hosta Walk (pp. 142–4), hostas are used throughout the garden. In the north-east corner at the foot of a wall facing east and beneath the boughs of ancient oaks, a sequence of peat beds is being made to accommodate many of the very small hostas, *H. venusta* for example, in its varying forms, in order to demonstrate how variable this species can be. The beds are being made here in sweeping, undulating curves, creating bays and bulges to give as long an area of peat wall as possible, since many dwarf hostas are stoloniferous and will grow to perfection in peat blocks, their stolons holding the wall together. Among others, the author will use *H. longipes*, some of the *H. sieboldii* forms including *shiro-kabitan* which must have total shade, and the cultivars *H.* 'Bobbin' and 'Po Po' which might be swamped in a normal border.

The hostas will be given further shade and shelter by being grown among dwarf rhododendrons, cassiopes, primulas and other peat-loving subjects, as well as hardy pleiones. The peat beds will need to be enriched as much as possible without upsetting the nutritional requirements of other plants, so the hostas will receive regular foliar feeds.

The beds will be sheltered from east winds and early sun by the grove of *Myrtus luma* on the corner of the Dell Garden, which itself is designed, rather like a doline, to lead one down a steep path between rock-retaining walls, on the top of which are to be grown at eye-level hostas valued for the whiteness of the underside of the leaves, notably *H. hypoleuca* and *H. pycnophylla*. *H.* 'Vera Verde' and *H.* 'Neat Splash' would grow in the crevices between the rocks.

Most of the Tardiana group will be grown in the old rose borders alongside the rose pergola in the centre of the garden. These glaucous hostas with small leaves associate particularly well with the blue-toned old roses, campanulas and lilies. Other hostas, especially those with white flowers, will be grown in the white garden.

Apart from the part of the garden near the western boundary there is still not enough shade, but this will come in time from the structures and plantings now in hand. Meanwhile we shall be watering and mulching the hostas to ensure that they remain in good condition – good enough to be worthy of the designation, National Reference Collection.

Minnesota Arboretum

The United States has its own National Hosta Display Garden, although unofficial, run under the auspices of the American Hosta Society, but originally the brainchild of Dr Leon Snyder, founder of the Arboretum. It is at the University of Minnesota's Landscape Arboretum at Chanhassan, Minnesota. The University of Minnesota is the site of the International Registration Authority for *Hosta*, the records of which are kept by Mervin Eisel, the registrar.

It was one of Dr Snyder's dreams that the institution should not only evaluate and display hardy plants, but should also help to educate the gardening public on how plants could be used in a landscape setting. Thus a number of herbaceous plantings evolved over the years alongside the trees and shrubs.

Although the Hosta Display Garden as such was started in 1975 hostas have been growing in the Arboretum for nearly 30 years. The first main site for the Hosta Display Garden was near the wildflower garden under a canopy of elms, and another small planting developed along the northern edge of a woodland pond. The Display Garden started out with 60 hostas and over the years, thanks to contributions from local hosta collector Dick Lehman and nurseryman and hybridizer, Bob Savory, and other dedicated members of The American Hosta Society, the Display Garden now houses well over 250 different species and cultivars, and today it is believed to be the largest public collection of hostas in the world.

In 1979 major changes took place regarding the future design and planting of the Arboretum and a new site near the new rhododendron display garden was chosen. It was completed in 1980 just before the American Hosta Society's annual convention which, that year, was held in Minnesota.

The hostas are planted beneath a canopy of wild sugar maples. The soil is a heavy clay with some leaf-mould. The shade and root competition has not presented a problem as far as the hostas are concerned because of the addition to the soil of peat moss, well-rotted horse manure and more leaf-mould, and a regular programme of feeding and watering. Some of the accompanying shade-loving plants have not been so eager to settle in but it is hoped that with the aid of the newly installed irrigation system they will adapt in time.

Low, white Lannon stone walls were constructed along either side of the main path giving continuity to the entire planting. The shape and size of the beds was influenced by the presence of the existing trees and topography. The hostas are grouped into beds by foliage type so that people can make side-by-side comparisons of them. The centre of the landscape bed is mounded up in order to create visual barriers and is planted with yew, Korean boxwood, *Hydrangea* 'Annabel', *Azalea* 'Northern Lights' and blue beech, and the herbaceous material includes dicentra, astilbe, bergenia, epimedium, pulmonaria, *Helleborus niger*, selected ferns and wild flowers and some shade-loving lilies. Large masses of single varieties of hosta have also been used at strategic places throughout the garden. It is felt that this wide range of woody and herbaceous plant material growing with the hostas creates more interest and serves as an educational tool for teaching the public about plants which can be grown together in the shade.

The Minnesota Arboretum library (The Anderson Library), now houses most of the records made by Mrs Frances Williams about her work on the genus *Hosta*. These records were donated by her daughter Miss Constance Williams, and are available for inspection by members of the public. Also on display are two large volumes of photographs of the hostas Mrs Williams grew in her own garden, and a large collection of slides (pp. 52–4).

Key to hosta types and their uses

This list is by no means comprehensive. There are now well over 1,000 cultivars and it is only possible to list the best in each category:

LARGE LEAVES
Clump up to 3 ft 6 in (106 cm) long
Leaf up to 1 ft 6 in (45 cm) long

Big Daddy	Blue Umbrellas
Blue Angel	Blue Mammoth
Blue Piecrust	Bressingham Blue

elata
Frances Williams
Green Acres
King Michael
Krossa Regal
montana & forms

nigrescens & forms
Northern Halo
sieboldiana & forms
Snowden
Sum and Substance

NARROW LEAVES:
Leaf up to 1½ in (4 cm) wide

Anne Arrett
Bold Ribbons
Caput-avis
Chartreuse Wiggles
Emerald Isle
Ginko Craig
Golden Isle
gracillima
Hadspen Hawk
Hadspen Heron
helonioides & forms
Kabitan

lancifolia & forms
longissima & forms
Louisa
rohdeifolia
Sea Sprite
sieboldii & forms
Subcrocea
tardiflora
tortifrons
Vera Verde
Weihenstephan
Wogon Gold

MEDIUM/LARGE LEAVES:
Clump up to 2 ft 6 in (76 cm) long
Leaf up to 10 in (26 cm) long

Antioch
Aspen Gold
August Moon
Blue Seer
Carol
crispula
Ellerbroek
erromena
fortunei & forms
Francee
Fringe Benefit
Gold Standard
Golden Sunburst
Honeybells
hypoleuca
Julie Morss

Lunar Eclipse
Moonglow
Moonlight
North Hills
Piedmont Gold
plantaginea
Phyllis Campbell
rectifolia
Royal Standard
Shelleys
Sun Power
ventricosa & forms
Wide Brim

MEDIUM/SMALL LEAVES:
Leaf up to 6 in (15 cm) long

Birchwood Parkys Gold
Blue Dimples
Blue Skies
Blue Wedgwood
Buckshaw Blue
capitata
Golden Medallion
Golden Sceptre
Golden Tiara

Hadspen Blue
Halcyon
Minnie Klopping
nakaiana
Pastures New
Pearl Lake
Platinum Tiara
tokudama & forms

SMALL LEAVES:
Up to 4 in (10 cm) long

Blue Moon
Chelsea Babe
Gold Edger
Golden Prayers
Hydon Sunset
kikutii (some forms)
Lemon Lime

Little Aurora
minor
Saishu Jima
Resonance
Sunset
Vanilla Cream
Venusta 'Variegated'

WHITE-BACKED LEAVES:

fluctuans 'Variegated'
hypoleuca

longipes 'Hypoglauca'
pycnophylla

LEAVES WITHSTANDING FULL SUN:

Blue Umbrellas
Francee
Golden Medallion
Golden Prayers
Golden Sculpture
Green Fountain
Honeybells
Invincible

Midas Touch
Moonglow
plantaginea
Royal Standard
Shining Tot
Sugar and Cream
Sum and Substance
Zounds

LEAVES WITH HEAVY SUBSTANCE:

Aspen Gold
Frances Williams
Gold Regal
Golden Medallion
Golden Sunburst
Invincible
Krossa Regal
rupifraga

sieboldiana & forms
Shining Tot
Snowden
Sum and Substance
tardiflora
tokudama & forms
Zounds

DWARF LEAVES:
Up to 2 in (5 cm) long

Bobbin
Po Po
pulchella
Shining Tot

Thumb Nail
Tiny Tears
tsushimensis
venusta

LEAVES WITH UNDULATE MARGINS:

Blue Piecrust
Bold Ruffles
capitata
Chartreuse Wiggles
crispula
elata (some forms)
fluctuans
fluctuans 'Variegated'
Green Piecrust
Grey Piecrust
Kabitan

pycnophylla
Regal Ruffles
Ruffles
Sea Drift
Shelleys
Subcrocea
Sun Power
The Twister
tortifrons
undulata

CORRUGATED LEAVES:

Aspen Gold
Big Daddy
Blue Seer
Bressingham Blue
Frances Williams
fortunei 'Rugosa'

George Smith
Love Pat
Northern Halo
sieboldiana 'Elegans'
tokudama & forms
Zounds

EPIPHYTIC:

hypoleuca
longipes
okomotoi

BLUE LEAVES:

Big Daddy
Big Mama
Blue Boy
Blue Moon
Blue Piecrust
Blue Seer
Blue Skies
Bressingham Blue
Buckshaw Blue
Hadspen Blue
Halycon
Love Pat
sieboldiana 'Elegans'
sieboldiana 'Mira'
tokudama

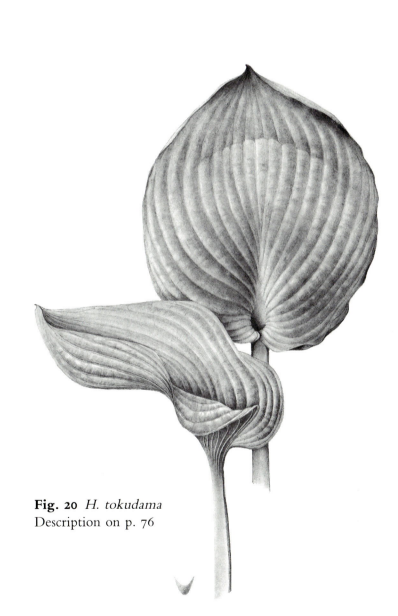

Fig. 20 *H. tokudama*
Description on p. 76

GREEN LEAVES:

County Park
decorata f. *normalis*
Emerald Skies
Erromena
fortunei 'Obscura'
fortunei 'Stenantha'
Green Acres
Green Fountain
Green Wedge
Invincible

kikutii & forms
lancifolia
longissima
Purple Profusion
rectifolia
rupifraga
sieboldii 'Alba'
tardiflora
undulata var. *erromena*
ventricosa

CREAM/WHITE-MARGINED LEAVES:

Antioch
Ellerbroek
fluctuans 'Variegated'
fortunei 'Hyacinthina
    Variegata'
Goldbrook
helonioides 'Albo-picta'
Moorheim
Neat Splash

Northern Halo
Resonance
Shade Fanfare
Spinners
Sugar and Cream
ventricosa 'Variegata'
Vera Verde
Wide Brim

GOLD/YELLOW LEAVES:

Aspen Gold
August Moon
Birchwood Parkys Gold
Gold Edger
Golden Medallion
Golden Prayers
Golden Sceptre
Golden Sunburst
Golden Waffles
Kasseler Gold
Little Aurora

Midas Touch
Piedmont Gold
Semperaurea
Subcrocea
Sum and Substance
Sun Power
Sunset
Vanilla Cream
Wogon Gold
Zounds

ROUND LEAVES:

Bengee
Big Daddy
Blue Umbrellas
Love Pat

Sea Lotus Leaf
Sum and Substance
tokudama
tokudama 'Aureo-nebulosa'

WHITE-MARGINED LEAVES:

Carol
crispula
decorata
Emerald Isle
fortunei 'Albo-marginata'
Francee
Ginko Craig

Gloriosa
Ground Master
Louisa
North Hills
sieboldii
undulata 'Albo-marginata
    (Thomas Hogg)

YELLOW/GOLD-MARGINED LEAVES:

Butter Rim
fortunei 'Aureo-marginata'
Frances Williams
Fringe Benefit
Golden Tiara
Green Gold

montana 'Aureo Marginata'
opipara
Shelleys
tokudama 'Flavo-circinalis'
Viette's Yellow Edge
Yellow Splash

LEAVES WITH GOLD CENTRE/WHITE MARGIN:

Anne Arrett
Lunar Eclipse
May T. Watts

Moonglow
Moonlight
Platinum Tiara

GOLD OR VARIEGATED LEAVES TURNING GREEN:

Chelsea Babe
Chinese Sunrise
Elizabeth Campbell
fortunei 'Albo-picta'
fortunei 'Aurea'
fortunei 'Hyacinthina Variegata'
Gold Haze

Gold Leaf
Hadspen Samphire
Nancy Lindsay
Oriana
Phyllis Campbell
Sharmon
ventricosa 'Aureo-maculata'

STREAKED/SPLASHED VARIEGATION:

Flamboyant	Sweet Standard
Neat Splash	*venusta* 'Variegated'
Northern Mist	Yellow Splash

LEAVES WITH GOLD/CREAM WHITE CENTRE/GREEN MARGIN:

Celebration	Reversed
Chelsea Ore	Sea Sprite
Chinese Sunrise	September Sun
Elizabeth Campbell	Sharmon
fortunei 'Albo-picta'	Shiro-kabitan
Gold Standard	Thea
Janet	*undulata* & forms
Kabitan	White Christmas
Phyllis Campbell	

LEAVES WITH UNSTABLE VARIEGATION:

Beatrice	Frances Williams
Flamboyant	Neat Splash
fortunei 'Albo-marginata'	Spinners
fortunei 'Hyacinthina Variegata'	Thea
	undulata & forms

SPECIMEN CLUMPS:

Borwick Beauty	*montana* 'Aureo-marginata'
fluctuans 'Variegated'	*nigrescens*
Frances Williams	Northern Halo
Fringe Benefit	*sieboldiana* 'Elegans'
Green Acres	*sieboldiana* var. *mira*
King Michael	Zounds

FOR EDGING:

Ginko Craig	*tardiflora*
Gold Edger	*undulata*
longissima 'Brevifolia'	Vanilla Cream

Fig. 21 *H. crispula*
Description on p. 72

FIRST LEAVES TO SPROUT:

crispula   Hadspen Heron
montana 'Aureo-marginata'

PEAT BED/ROCK GARDEN:

Anne Arrett   Little Aurora
Blue Moon   *minor*
Chartreuse Wiggles   Sea Sprite
Ginko Craig   Tiny Tears
gracillima   *tortifrons*
Hydon Sunset   Vanilla Cream
kikutii 'Polyneuron'   *venusta*
Lemon Lime   Vera Verde

LANDSCAPING LARGE AREAS:

'Erromena'   Royal Standard
Honeybells   Tardiana hybrids
North Hills

GROUND COVER:

clausa   Groundmaster
decorata   *longipes*
Francee   *venusta*

FIRST TO BLOOM:

capitata   *nakaina*
crispula   *undulata*

FRAGRANT FLOWERS:

Honeybells   Sugar and Cream
Invincible   Summer Fragrance
plantaginea & forms   Sweet Standard
Royal Standard   Sweet Susan

DUSK BLOOMING:

plantaginea & forms

LATE TO BLOOM:

gracillima	*rupifraga*
kikutii 'Caput-avis'	*tardiflora*
longipes 'Aequinoctiiantha'	*tardiva*
plantaginea	*tortifrons*

WHITE FLOWERS:

Bright Glow	sieboldiana 'Elegans Alba'
Butter Rim	*sieboldii* 'Alba'
Emerald Isle	Snowflake
Louisa	Tinkerbell
plantaginea & forms	Weihenstephen

BLOOMS NEVER OPENING:

clausa

6.

CULTIVATION OF HOSTAS

When to plant

The best season to plant hostas is early spring, just as the new shoots are emerging. This enables them to establish themselves as soil and weather warm up. However, some growers recommend planting, or transplanting, in late summer or early autumn, and this can be just as successful provided that there has been plenty of rain during the summer to keep the ground sufficiently moist. The hostas will then have enough time to re-establish their root systems before the first of the frosts.

The worst possible thing to do is to plant in the dead of winter when the hosta is totally dormant and the ground at its coldest. In such conditions roots could rot, and small divisions which would succeed if planted into warmer ground in spring are particularly vulnerable. This applies especially to some of the less vigorous sorts such as *H.* 'Kabitan', *H.* 'Louisa', *H.* 'Butter Rim' and *H. decorata*. These observations apply in Britain where summers are not hot enough for some hostas ever to reach their full potential and vigour.

It is always wise to pot up any rare, valuable or slightly suspect hostas and over-winter them in a cool greenhouse, planting them out in spring if they are big enough. In any case, some varieties grow away best if started in pots and grown on for at least a year. Such treatment pays dividends with *H.* 'Kabitan', *H.* 'Louisa', *H.* 'Butter Rim', *H. decorata* and with others like *H.* 'Janet', *H.* 'Golden Isle' and *H.* 'Ginko Craig' which then grow away far better in British gardens, than if they were set directly into the ground, however good the soil.

Spring planting has another advantage. If a newly planted hosta turns out to be in the wrong place, it can easily be lifted and re-planted without putting it under undue stress. The same cannot be said for autumn-planted hostas as they disappear from sight once the leaves have died.

Hostas take anything up to five years in the same place to reach maturity, so the more often they are moved, the longer this will take. Most hostas also exhibit juvenile and adult phases, the leaves of the juvenile phase being usually smaller and narrower than those on a mature plant. Frequent moving could keep a hosta permanently in its juvenile phase. Of course there are occasions when one needs to move a hosta, and even move it in full leaf: it is often the most convenient and practical time to give away and exchange plants. If a hosta is to be lifted in full leaf or even in full flower – and this can be done successfully – it is essential that it is lifted with enough root ball, and that this operation is not carried out in extreme weather, particularly in times of drought or in blazing sunshine.

The clump should be moved as soon as possible to a well-prepared planting hole, carefully lowered in, firmed and then given enough water to speed its recovery. The clumps should be covered in damp sacking if re-planting has to be delayed for any reason. Damage usually occurs only when one is trying to move hostas that are too big to handle with comfort. If wilting occurs it is worth sacrificing the current season's leaves by applying a continuous fine spray of water during hot weather as well as applying water directly to the root ball. If wilting persists, it is worth reducing the length of some or all the leaves by half.

How to plant

The technique of planting a hosta depends mainly on its expected size at maturity.

For large hostas like *H.* 'Sum & Substance', *H.* 'Krossa Regal', *H. fluctuans* 'Variegated' or *H. sieboldiana* and its forms, it is necessary to dig a hole as large as that needed for a shrub, that is 3 ft (1 m.) across and 18 in (45 cm) deep. The bottom of this hole should then be filled with a 4–6 in (10–14 cm) layer of well-rotted farmyard manure or, failing that, garden compost, and then covered with a 3 in (8 cm) layer of earth mixed with fine leaf-mould. It should then be carefully firmed and the hole back-filled to the level of the border. The hosta's roots should not come into direct contact with the manure, since it could cause the new season's leaves to scorch, particularly if the plant is exposed to bright sunlight. Excessive manure at the roots damages the cell structure of the leaves as they are developing. If possible this preparatory work should be done a month or more before planting time so that the ground has a chance to settle. At planting time all that need be done is to dig out a hole big enough to take the hosta comfortably. A loose mound of soil should be piled up centrally at the bottom of the hole and the roots of the hosta shaken from its potting compost and arranged over this mound with the crown resting on the top. If the hosta's root ball is pot-bound, then mound planting will not be practicable and the firm root ball should be so placed in the planting hole that its crown is level with the surface of the border. The plant should have earth packed round it and be well-firmed, treading it down with the heel if necessary. Finally the hosta must be thoroughly watered in, using a hose trickling gently over the planting hole for several hours. Newly planted large hostas will need a gentle trickle of water every day for at least a week, preferably in the early morning and evening. Watering can then gradually be reduced, depending upon the amount of sunlight and natural rainfall. Turgid leaves will indicate that all is well.

For smaller hostas (the Tardiana group, *H. undulata*), a hole half the size would be ample, and it is not necessary to use as much manure provided the border was adequately manured originally, though some is preferable. Again, leaf-mould and moss peat should be incorporated with the soil which is returned to the planting hole. This promotes the development of a strong root system and that in turn produces a balanced plant.

For dwarf hostas (*H. venusta, H.* 'Po Po, *H.* 'Blue Moon'), the size of the planting hole seems to be less important than the provision of a suitable rooting medium: equal parts of peat, leaf-mould and loam. Many of these tiny hostas are stoloniferous, and a made-up bed 6 in (15 cm) deep but a yard (1 m) or so square may be preferable to a small planting hole in ordinary soil.

Smaller hostas need less water to become safely established and, again, after an initial soaking, amounts and frequency will depend on the size of the hosta and on prevailing weather conditions.

Newly transplanted hostas will become established more quickly if the flower spikes are removed.

Whichever size of hosta has been planted, surround it with a ring of mulching material (not farmyard manure for the smaller ones), to keep the rays of the sun off the hosta's roots, to conserve moisture and to suppress weeds. The depth of the mulch should be appropriate to the size of the hosta: for large hostas a mulch of 3–4 in (7–9 cm) deep is perfect, but for smaller hostas it should be proportionately smaller – scarcely half an inch (1 cm) for the smallest.

Planting conditions

Hostas are more generally grown for the magnificence of their leaves than for their flowers, so cultivation is usually geared to the production of large clumps of unblemished foliage.

To reach the peak of perfection, hostas need a moist, fertile soil and adequate shade and shelter.

The ideal soil for hostas is a friable loam, just on the acid side of neutral (a pH of about 6), with an abundance of organic matter. A high proportion of chalk in the soil will give glaucous-leafed hostas a muddy tinge, but they can be grown in an alkaline soil quite adequately provided there is sufficient organic matter.

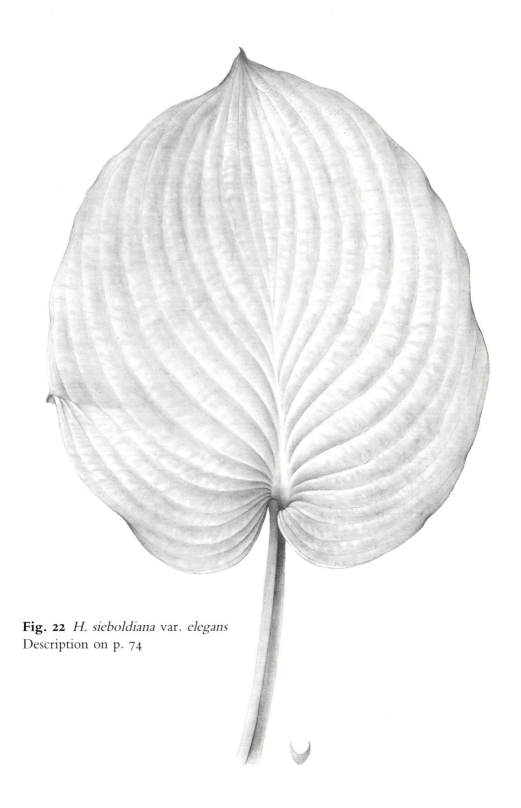

Fig. 22 *H. sieboldiana* var. *elegans*
Description on p. 74

Feeding

There is probably no other group of ornamental plants that responds so rewardingly to feeding as do hostas. They will amply repay the time and trouble it takes to incorporate manure into the ground before planting as well as regular applications of granular fertilizer or foliar feed.

Organic manures

If a border is being prepared for hostas, the ground should be double-dug and manure added to the bottom of the trenches several months prior to planting, otherwise the manure should be applied directly to the planting holes. If poultry manure is used (and it is particularly high in nitrogen), it is important that it is first mixed thoroughly with sawdust, in the proportions of one part manure to two parts sawdust, unless it is obtained as old-fashioned, strawy poultry manure.

Garden compost may also be used as a manure. If well-made and mature, it is one of the best and most balanced of organic materials to dig into the ground.

Reconstituted chicken and cow manure have become popular substitutes for manure *au naturel* and they are much easier and more convenient to handle.

Liquid manures

Liquid manures afford a useful means of supplying vital nutrients to hostas growing in pots or tubs, as well as for supplementing the nutrients available to hostas growing in open ground. Liquid manure must not come into direct contact with the hosta leaves as it will discolour and probably burn them. It should be applied to the soil only, which should be moist already, so that it can properly absorb this highly concentrated plant food.

Liquid manure should be diluted to half-strength if applied to young or freshly divided plants.

Mulches

Mulching is essential to the sucessful cultivation of hostas. They need a regular regimen of spring and autumn mulches if they are to become really sumptuous. The autumn mulch should be of well-rotted farmyard manure one year and leaf-mould or garden compost the next, so that too high a concentration of nitrogen is not accumulated in the soil. The autumn mulch should be applied as soon as the last hosta leaves have died away and should be applied as described in the section 'How to plant' (p. 173). The mulch should not cover the hosta crowns which might rot. The crowns themselves can be protected against severe weather by a light litter of dry bracken fronds, straw or salt hay (US). These materials should be removed in early spring before growth begins, otherwise the emerging leaves can be distorted.

The autumn mulch acts as a blanket through the winter, ameliorating the extremes of temperature and preventing the ground round the hostas from alternately freezing and thawing, a process which is thought by many growers to cause crown rot. Equally important, it provides a supply of nitrogen ready for the greedy hosta roots when they demand it in the spring.

The spring mulch should primarily be of moisture-conserving fibrous materials with a relatively low nutrient content. Moss peat and leaf-mould are ideal, especially if mixed with the products of a well-balanced compost heap. The spring mulch should act as a soil conditioner, but if for any reason the autumn mulch was not applied, or the hostas show signs of needing nutriments in the early part of the summer, these are best supplied in the form of a quick-acting fertilizer. Foliar feeds are probably the best for use on hostas; many growers use tomato fertilizer with very good results, but undoubtedly the best are the reconstituted manures referred to previously.

Grass clippings are a useful addition to the compost heap if well-mixed with other organic materials when they will rot down properly but on no account should they be used raw as a mulch round hostas. They compact down, preventing rain from entering the ground and they heat up too much, to the detriment of the hostas.

Leaf-mould

After animal manure, leaf-mould is the next best mulching material for large hostas and always the best for smaller hostas, but it is important to remember that leaf-mould collected from trees growing in a chalky or alkaline soil may have too high a pH for acid-loving plants. If hostas are to be grown among rhododendrons and azaleas, then leaf-mould with a low pH will be required.

Fine leaf-mould, made from oak or beech leaves, is an excellent medium if mixed with a good quality loam, so should not be wasted on the general compost heap. Leaf-mould and compost should always be kept in separate bins since they are broken down by different sorts of bacteria.

Fertilizers

Fertilizers are essentially processed sources of nutrients. Their efficacy is expressed in terms of the proportions of N (nitrogen), K (potassium) and

P (phosphorus). Of these, nitrogen, which is vital for healthy stem and leaf growh, is the most important for hostas.

The larger-leafed hostas make enormous demands on the available soil nitrogen and if insufficient is available, optimum growth will not be achieved. April and May are the most critical times when lack of nitrogen can markedly check growth. At this time a quick-acting nitrogen fertilizer should be applied. However, excessive applications of nitrogen fertilizers will encourage lush, sappy growth which is more easily damaged by the elements and it is also more tempting to slugs and snails. To counter-balance this potassium is needed as this helps to produce firm leaves, and increases a plant's resistance to disease. It is used by hostas in large quantities.

P, phosphorus, is needed to build healthy roots and make rapid growth in spring. Most garden fertilizers contain N, P and K. Balanced fertilizers contain them in roughly equal proportions, but in high-nitrogen fertilizers there may be twice as much nitrogen as phosphorus and potassium.

Foliar feeds

When hostas are in full leaf at the height of the growing season, there may well be no earth visible between the plants and in this case the best way of applying quick-acting fertilizer is a foliar feed. Such a feed should only be given in the cool of the evening otherwise the leaves may scorch; it should only be applied during dry weather as rain would wash the liquid off the leaves. To make it adhere better a few drops of washing-up liquid can be added to the watering can. Unless it is absolutely essential, foliar feed should not be given to glaucous-leafed hostas as the pruinose coating can be damaged, producing a blotchy appearance.

Dried organic fertilizers

Although strictly organic, powdered forms of plant nutrients such as dried blood, bone-meal, hoof and horn and fish meal, are usually lumped together with fertilizers because their manner of application is similar. All of these can be used in the cultivation of hostas.

Watering

Hostas need plenty of moisture to sustain them although, paradoxically, large established clumps of heavy substance can tolerate periods of drought although they will not thrive.

Many species grow in the wild in places where roots are actually in water, often in areas where the precipitation (rainfall, snow, fog and mist), depending on the season, is almost continuous. Hostas in the wild are, therefore, nearly always to be found shrouded in mist or low cloud, in diffuse sunlight with high humidity. Their leaves are always wet and the plants always have moisture at their roots. There are, of course, exceptions to this, *H. hypoleuca* being one.

These natural conditions cannot easily be replicated in gardens in the western hemisphere, but the aim should always be to keep the soil moist at all times, and to provide adequate shade and shelter. This is absolutely essential at the start of the growing season when the young leaves are developing.

The best way to water hostas (especially the glaucous-leafed types), is to apply water close to the stem of the plant, below the foliage, leaving the surrounding earth dry. All the water is then applied to the root zone where it can be used by the plant.

Water is best applied either by a can (without the rose) or by hose, but not through a nozzle. If it is absolutely necessary to use a sprinkler, leave it watering in one place for several hours.

Sprinklers should never be used on hostas in strong sunlight: the droplets resting on the leaves can focus the sun's rays as effectively as a lens and can cause severe scorching. Watering should always be done (especially when a sprinkler is used) on a cloudy day, or early in the morning. Evening watering tends to attract slugs and snails. Another good reason for watering early in the day is that the plants will have a full day to make use of the water which is absorbed in daylight.

Hostas grown in tubs, terracotta pots and other containers are particularly vulnerable to rapid moisture loss. Soil-less peat-based mixes should be avoided as a long-term planting medium for this type of hosta culture as the plants will dry out more quickly than if planted in a soil-based mix. Once dried out, re-wetting is a slow and unsatisfactory process.

Shade

Hostas are relatively new to cultivation in Europe and America and how they grow in the wild is still relevant to their needs in our gardens. Their range of wild habitats is almost as diverse as the hostas themselves, but they are seldom exposed to strong sunlight and even those growing in flat marshlands and valleys are surrounded by tall grasses, but with a little ingenuity we can give them acceptable growing conditions.

Shade is more than merely the absence of direct sunlight: it creates its own micro-climate of enhanced humidity and lowered temperature. It reduces the ambient temperature of the air as well as the temperature of the soil. As a result, plants growing in shade are cool, and operate at a reduced rate of transpiration, which in turn means they take up less moisture from the soil. Shade also reduces the rate of soil-moisture evaporation so that in general soil in shade will be more moist than soil in the sun.

The general rule is therefore to grow most hostas in shade or dappled shade; but there are exceptions. Current hybridizing programmes in America are producing hostas that will flourish in full sun, and there are now some varieties that will take the sun are far south as North Carolina and Georgia. *H.* 'Sum & Substance', with leaves of exceptionally heavy substance, and broad-leafed *H.* 'Golden Sculpture' will both withstand this unusual treatment, but they do of course still need plenty of moisture at their roots. *H. plantaginea* needs hot sunshine to produce its fragrant white flowers.

Many white-margined hostas scorch badly in full sun, even in more northern parts of Britain, *H. ventricosa* 'Variegata' is one such, as well as *H. crispula*, and *H. fortunei* 'Albo-marginata' positively does not flourish or exhibit its outstanding variegation if grown in any sun at all.

Similarly some small thin-leafed hostas like *H.* 'Kabitan', *H.* 'Wogon Gold' and *H.* 'Anne Arrett' scorch badly, as do some of the white-margined, gold-leafed varieties. *H.* 'Moonlight' comes to mind here, but *H.* 'Moonglow' and *H.* 'Lunar Eclipse', the same colouring but different parentage, seem impervious.

The exact amount of shade needed by different hostas varies considerably and can only be established by observation and experimentation, always erring on the side of too much, rather than too little shade. There are also substantial geographical variations. In the American mid-west, where light intensities are high and humidity is low, all hostas have to be grown in deep shade; in southern Britain they can be grown in partial shade – ideally the light, dappled shade cast by the leafy canopy of tall, deep-rooted oaks, as at the Savill Garden in Windsor Great Park – though many will flourish out in the open, provided there is sufficient moisture at their roots. In west coastal Scotland, where rainfall and humidity are high, and light intensities low, hostas can be grown with virtually no shade at all.

Lack of adequate shade affects different hostas in different ways. At the very least, it can make them look unsightly by causing the leaves to discolour. The glaucous-blue hostas, grown for this attractive characteristic, are affected most and some of these can even scorch (*H.* 'Hadspen Heron'). Similarly the leaves of some variegated hostas have been known to revert entirely to green when grown in full sun (presumably in self-protection), while others will scorch so badly that their leaf area becomes inadequate to sustain them. *H. undulata* falls into this category. Some may even die if exposed to full sun.

The punishing effect of the sun in such cases is the result of physical damage. Scorching reduces the area of leaf that can photosynthesize, so year by year the plant will diminish and its vitality will fall until it perishes.

Natural shade

Natural shade is variable as to its quality and one has to balance the benefits of the shade cast against the problems created by the trees that cast it.

The great advantage of natural shade is that trees can provide an attractive setting for hostas. Yet not all trees are equally desirable as shade trees for hostas. The period when they put on their leaves can vary by as much as two months – a third of the growing season. A birch coming into leaf in April may seem ahead of the season, but a catalpa, coming into leaf in June, would leaf too late to prevent the tender young hosta leaves growing beneath it from receiving a severe scorching. Equally, the sooner a tree puts on its leaves, the sooner its roots start to draw on the soil moisture reserves, and for this reason one should choose trees whose roots go deep, oaks for example, in preference to those with shallow roots, since they will also compete directly for the same nutrients as the hostas.

There are also great variations in the density of the shade cast. Which is best may depend upon the intensity of the sunlight in the locality where the hostas are to be grown.

Not all hostas are well-suited to planting under trees. *H. tokudama* and its forms and hybrids have cup-shaped leaves which collect the litter that falls from overhanging trees and become so badly stained and discoloured by rain-drip that they are hardly worth growing.

Artificial shade

The great advantage of artificial shade is that it can be controlled, not only in terms of its precise density, but also in terms of when and where it occurs.

The most basic sort of artificial shade is that cast by walls, both those of houses and outbuildings and those on their own. In the northern hemisphere the soil immediately beneath north or north-west facing walls receives very little sun, except at midsummer, and possibly none at all if the sun is also blocked by trees or other structures. Such a situation can make a most hospitable site for hostas. In British gardens this type of site is fairly sheltered, as the prevailing winds tend to be south-westerlies. The problem is that there will be a rain shadow, so constant attention must be paid to watering. The addition of large quantities of organic matter to the soil will also help to keep up the moisture content.

Shade houses

The ultimate in measurable, manageable, controllable shade is a shade house, and these structures are now much used by nurserymen producing hostas. They make as nearly perfect a place as is practicable in which to cultivate hostas, and just about indispensable for anyone growing hostas for showing and exhibition work and, possibly in the short term, for hostas being used for flower arrangement when the leaves must be in absolutely perfect condition. In Britain care must be taken not to make the shade too dense or the leaf colouring will be affected.

The old-fashioned wooden lath house, such as that used for growing hostas as Weihenstephan Trial Garden in Munich, has much to commend it for the ornamental garden, as has the wooden lath canopy protecting the hosta demonstration border at Cleave House in Devon, though for sheer efficiency and cheapness, it has been superseded by modern, plastic-covered structures, usually tunnels. One of these plastic shading materials actually apes the old laths, having strips of 2 in (5 cm) black plastic, 2 in (5 cm) apart. This arrangement is most useful on a square structure.

Other modern shading materials are woven and net-like plastics, usually available in green or black, and these are made in various grades of shading – 25 per cent, 50 per cent and even 75 per cent shade. These net tunnels are used by the majority of specialist hosta nurseries in Britain. 50 per cent seems ideal in southern England.

Shelter

The leaves of hostas, especially the larger-growing sorts, are easily damaged by wind, so hostas should never be planted in open, windswept places. Salt-spray, borne on the wind, is even more harmful. Thoughtful siting, with protection from prevailing winds, will usually reduce the severity of the problem considerably.

Wind damages not only by its sheer physical force – bruising the leaves, tearing and breaking them – but also by scorching, which is manifested as browning of the margins, similar in effect to sun scorch. Scorching particularly affects hostas with leaves of thin substance and those with variegated margins, while hostas of heavy substance are more prone to splitting or tearing.

Hostas should never be grown in a draught or a wind tunnel, which is usually created by wind funnelling through narrow gaps between buildings: office precincts and shopping malls are particularly notorious. A similar but less violent effect is produced by wind being funnelled down a valley when there are no obstacles to break its force.

It is sensible, therefore, not to plant hostas in an exposed site, but if this is unavoidable, then some sort of artificial shelter should be provided. Artificial wind shelter is usually taken to mean fencing of some kind, and the best fences for this purpose are made of modern flexible plastics and are semi-permeable – that is, they filter the wind, breaking its force.

Such measures should never be regarded as more than temporary. A dense hedge, either evergreen or deciduous, provided the bare branches are close-knit like hawthorn, provides the best form of protection against wind. A mixed border of shrubs is nearly as effective and probably more attractive if used as an internal garden boundary.

Plastic-netting shade-tunnels are equally effective as protection from the force of the wind for the plants inside them.

Propagation

Vegetative propagation

The easiest way of increasing hostas is to divide them. This is most effectively done with a mature clump by taking slices out of it as if it were a cake. If good soil is then used to fill in the hole from which the slice was removed, the plant will grow back into the new soil and continue to grow as if it had not been disturbed at all. The slice should be cut out with a sharp spade and the whole operation carried out in early spring just as the crowns are beginning to appear above the surface of the soil. There will obviously be some damage to the emerging crowns, but neither the main plant nor the segment removed will be permanently damaged. The segment can either be re-planted whole or divided into even smaller pieces. Any excessive or extra long roots should be cut off to encourage new root growth.

Smaller clumps can be dug up whole and split through the centre with a spade or sharp knife. Using such a knife, hostas can be split up into quite small pieces. Those hostas which form very big clumps, growing undisturbed for many years, generally form a hard root-core from which both roots and leaves emerge: a proportion of this root-core needs to be cut carefully away, and pieces of this core, with roots, will usually grow away even if there are no growth buds or eyes. Some hostas such as *H. venusta* and *H. clausa*, which are stoloniferous, grow quite differently; every fan of leaves grows from its own small root-core and a whole individual plant can be shaken or teased apart very easily. Either way, newly-divided hostas need to be treated as invalids and cossetted accordingly. It is often advisable to grow small divisions in pots filled with a proprietory compost, containing a slow-release fertilizer, before setting them out in the ground.

Some hostas are extremely slow to increase; *H. tokudama* and its forms, and *H. montana* 'Aureo-marginata', for example. A method has been devised by which such hostas can be encouraged to increase more quickly. This is known as the 'Ross Method' and was originated by Henry A. Ross of the Gardenview Horticultural Park, Strongsville, Ohio, USA. It is based on the observation that hostas will mend damaged tissue by callusing and will then make new growth from the callus. Plants have the ability to send out roots, as well as buds and shoots from an injured surface. Hostas naturally produce buds at the base of the crown and when an incision is made the healing process forms a callus over the damaged tissue which stimulates the production of the new growth buds.

The method is used in spring when the leaf buds are quite advanced. The technique is first to scrape back the earth from round the crown of the hosta, revealing not only the buds but also the firmer flesh of the root-stock (from which the roots spring), and then to insert a narrow, sharp knife into the bud just above the basal plate where the bud merges with the rootstock. The knife is then pressed downwards right through the rootstock. A second cut can be made at right angles to the first, quartering the rootstock below the bud. Each bud can be dealt with in this way but it is much easier to make these cuts on young plants with only a few crowns than on mature clumps. The cuts should be dusted with a fungicide powder and the earth firmed back round the plant. Each cut section of rootstock will callus and produce a new plant next spring (four plants instead of one where the stock has been quartered). Meanwhile the current season's leaves will grow on as if undisturbed, apart from a slight yellowing.

There are a number of variations on this method of propagation, such as cutting off some of the leaves and emerging flower spikes at the base of the plant, when buds or 'eyes' will appear on the surface of the wound, but the basic idea is the same. Propagation by these methods requires a steady hand but they are used successfully by both amateurs and professionals alike.

Growing hostas from seed

The seed of nearly all hostas germinates readily and produces an abundance of plants. However, hostas interbreed so freely that the seedlings cannot be relied upon to replicate at maturity the parent from which the seed was gathered, whether a species or a cultivar. Even when seedlings appear similar to the parent, they cannot be called by the same name because they are almost certainly genetically different.

The only species which comes true from seed is *H. ventricosa*, which is apomictic. The seedlings of all other species, whether deliberately crossed or not, must be considered hybrids. They are not usually worth dignifying with a cultivar name and it is totally misleading to pass them off as a species, a practice which is far too frequent and which causes endless problems of identification. Frankly, most seedling hostas are not worth growing.

However, raising one's own plants from seed can be a satisfying hobby if crosses are made deliberately between carefully selected parents, chosen in the hope of producing a particular style of hosta. It is the results of such well-chosen crosses that have made hostas some of the foremost perennial garden plants.

Gathering seed

The optimum time for gathering the capsules is just as they start to turn yellow. Once gathered, they should be left to dry on blotting paper laid flat in a sunny place. When the capsules start to split, the seeds should be shaken out and cleaned. This means sorting the black, shiny $\frac{1}{4}$ in (6 mm) long, cylindrical, fertile seeds from the pale, buff-coloured infertile seeds and the chaff. Finally, the seeds should be put through a wire sieve to remove the remaining debris.

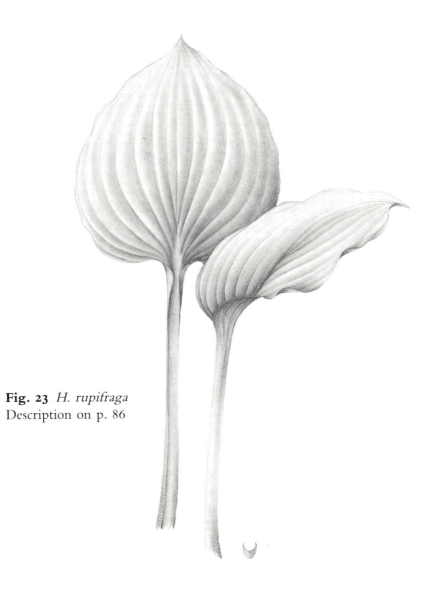

Fig. 23 *H. rupifraga*
Description on p. 86

Hosta seed is best sown as soon as it is ready since there is no period of dormancy. It tends to lose its viability after about a year if stored under home conditions, though if kept under scientifically controlled conditions it should remain viable for decades.

If seed is sown in a gently heated greenhouse or in a propagating unit it will usually germinate at once. If it is sown in open ground in autumn it will not germinate until late spring. The higher the temperature, the faster the seed will germinate, although a temperature of 79°F (26°C) would seem to be the optimum, and excessive heat retards germination. The speed of germination varies from species to species, but one to three weeks is about normal.

Sowing the seed

Hosta seed should be sown in sterilized pots or trays (flats) containing a compost which is loose, friable and porous. A half-and-half mixture of garden loam and sieved leaf-mould or peat, with a little sand added, will give good results. Hosta seed can be raised in commercially prepared mixes; on the whole a soil-based mix is preferable. There is a suggestion that seed may rot more readily in peat-based mixes, but this seems to be more a matter of opinion than fact.

Seed should be sown thinly since virtually every seed will germinate. It should be covered with its own thickness of sowing mix or with sharp, horticultural grit. When four leaves appear, each seedling should be transferred to a 2 in (5 cm) peat pot containing the same mixture as that used for the seeds with some added grit or vermiculite to retain moisture, and to prevent 'damping off'. The seedlings should be fed three or four times during the growing season with liquid fertilizer diluted to half strength. At the beginning of the following growing season the seedlings can either be transferred to larger plastic pots or planted in the ground in nursery rows after being hardened off. The strength of the liquid feed should be increased to three-quarters the strength of that used on mature plants.

Micropropagation

Tissue culture is merely an extension of the propagating techniques with which most gardeners are familiar. Plants which increase easily by traditional means increase easily in tissue culture.

In tissue culture one takes a micro division (instead of a normal division) and inserts it in agar jelly inside a test tube (instead of potting compost).

In theory tissue culture is something the layman can carry out at home. In practice the sterile conditions needed, and the necessity of providing an environment in which both light and temperature are controlled precisely, are too inconvenient and difficult to attain.

The great virtue of tissue culture is its capacity for bulking up stocks. Ten thousand units from one plant cell in a matter of months is now standard practice, and this is ideal for professional nurserymen.

Micropropagation has far-reaching implications for hostas because of the frequency with which new varients arise and because of the instability of some of these. Indeed, many of the most desirable new hostas have arisen in this way.

7.

PESTS AND DISEASES

Pests

By and large hostas are good, healthy, permanent perennials and remarkably free from pests and diseases. There are, however, two pests which can cause absolute havoc. These are slugs and snails. The impact of other pests is trivial by comparison.

Slugs and snails

There is no ultimate solution to the problem of slugs and snails, but there are some measures that can be taken to avoid making the garden a veritable breeding ground for them. The first essential is garden hygiene. This entails the discipline of not leaving decaying plant-remains lying about. It is tempting to leave frost-killed hosta leaves on the borders to rot down into the soil but it is far better to put them on to the compost heap, as if left on the soil they provide not only perfect sheltering places for these pests but also ideal breeding grounds. One also has to consider whether or not to apply mulches or top dressings of garden compost or farmyard manure, since these materials consist of little else than decaying vegetable remains. Gardens in which no organic mulches or manures are used have smaller populations of slugs and snails than gardens in which organic manures and fertilizers are used regularly.

In practice one grows the best hostas if one uses organic materials to promote growth, and chemicals to destroy the slugs and snails. The most commonly used chemical deterrent is metaldehyde, which is usually mixed with bran or some other material like molasses which attracts the slugs and snails. It works on the basis of anaesthetizing the pests so that they dehydrate and die if the weather is dry or lie drugged and vulnerable to their natural predators, but in moist and wet conditions they can recover. The air can be particularly humid under the leaves of large hostas and this makes hosta clumps a particularly inviting recovery area for slugs and snails. Methiocarb is a newer product, and more effective in that it has a more direct toxic action, making it impossible for the slugs and snails to recover. Although metaldehyde can be mixed with bran at home and placed on the borders in small heaps, it is usually now sold as pellets which are much more easily applied. Methiocarb is also supplied in pellet form. These pellets are nowadays turquoise-blue in colour since the colour blue is unattractive to birds. The pellets also contain a substance which makes them less attractive to domestic pets.

Both products work by forcing the slug or snail to produce excessive slime. The pellets remain effective for only three to four days, and repeat applications are essential, especially after rain. 2 lb (1 kg) of pellets should treat 2 acres (1 hectare). If the pellets are scattered sparingly on the ground and not piled into little mounds, they will be unlikely to attract or be noticed by birds or domestic pets.

An alternative remedy which is becoming more popular nowadays as we become more aware of the damage that powerful chemicals can do to the environment in the long term, is application in liquid form of products based on aluminium sulphate, and these are less expensive than methiocarb or metaldehyde. They are reasonably effective if watered directly on to the hosta leaves and the surrounding earth and if regularly re-applied, particularly after rain, and they are relatively non-toxic to birds and pets.

There are various other, less toxic, means of disposing of slugs and snails which have their advocates. One such is the 'slug pub'. Beer is poured into a deep plastic dish, with a wide slatted lid, which is sunk into the ground. The slugs are attracted to the sweet, intoxicating smell, fall in and drown. One then has to deal with the remains. Another method is to place upturned halved grapefruit skins near the hosta clumps. The slugs will venture into these but one is still faced with the problem of what to do next since the slugs are still alive. A garden full of upturned grapefruit skins is, of course, very unsightly. Crushed eggshells or coal ashes scattered in a ring round the hosta clumps, or even soot, have also been tried with varying success. Salt is lethal to slugs, but it is also toxic to plants.

A product now being tested in America is reported to be effective. It is a thickened water system containing 4 per cent metaldehyde, which is applied to the ground by squeezing it from a plastic bottle in small drops or in a line. The manufacturers claim that it is highly effective because it contains a selective chemical attractant which is far more powerful than the smell from the plants on which the slugs and snails normally feed. They die a kinder death, it is claimed, as it also contains a synergist for the metaldehyde, resulting in immediate death. The manufacturers of this product claim that pellets work so slowly in the slugs' digestive system that they can and do sometimes overcome the effects of the poison by excretion. The toxin in this product cannot be excreted so the slugs cannot survive. It is effective on the ground for three to four weeks and is then washed into the soil as carbon dioxide (which is beneficial to the plants). The high cost has prevented its being marketed on a large scale in Britain but it is available packaged in small quantities suitable for use on pots and peat beds.

A product which has been available in Britain and many other countries,

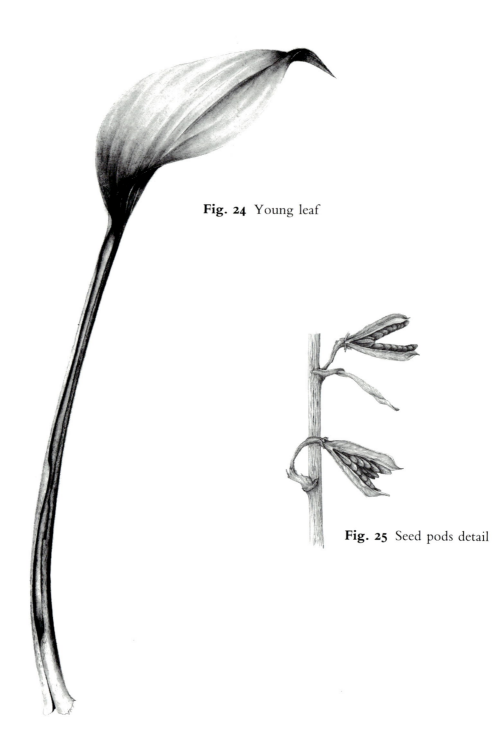

Fig. 24 Young leaf

Fig. 25 Seed pods detail

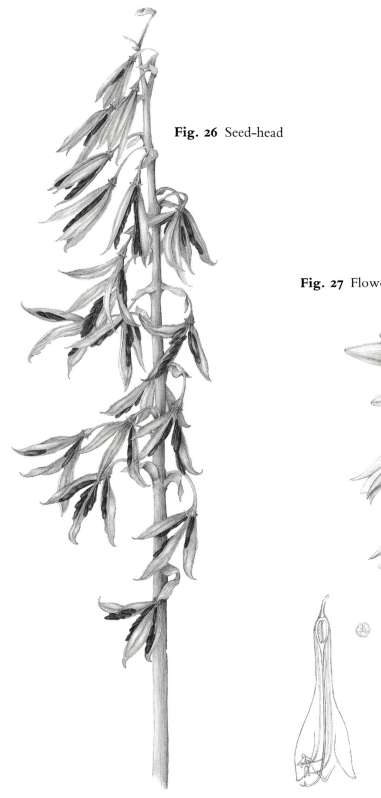

Fig. 26 Seed–head

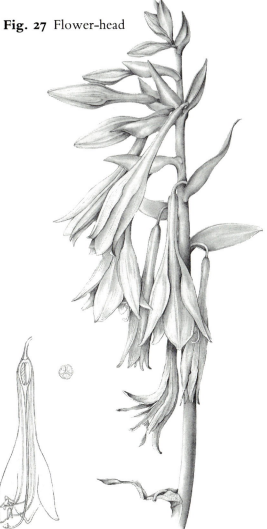

Fig. 27 Flower–head

and which has satisfactorily undergone exhaustive tests, is impregnated tape. This is a continuous barrier of paper-based tape filled with a specially formulated bait containing a small percentage of metaldehyde, which is laid in or on the soil. The manufacturers claim that it is clean and safe to handle and does not attract wildlife or pets, and is so effective that in dry conditions it is recommended that the area should be watered to encourage the slugs to feed. The tape comes in reels and is applied in long or short lengths; a triangle of it can be placed round each hosta clump. Slugs and snails eat holes in the tape so that it is possible to see exactly where infestations are heaviest and extra lengths can be used as required. No tape need be wasted because uneaten tape can be taken up and re-laid on other parts of the garden. Any residue can be hoed into the ground where it will continue to work on those slugs which live under the surface of the soil, or it will simply rot away.

HOSTAS MOST VULNERABLE TO SLUG AND SNAIL DAMAGE

H. 'Anne Arrett'	H. 'Hadspen Samphire'
H. 'August Moon'	H. hypoleuca
H. 'Birchwood Parkys Gold'	H. 'Kabitan'
H. 'Chartreuse Wiggles'	H. kikutii
H. crispula	H. pycnophylla
H. fortunei 'Albo-marginata'	H. 'Sea Sprite'
	H. 'Shiro-kabitan'
H. fortunei 'Albo-picta'	H. subcrocea
H. fortunei 'Aurea'	H. undulata & forms
H. 'Francee'	H. ventricosa & forms

HOSTAS MOST RESISTANT TO SLUG AND SNAIL DAMAGE

H. 'Gold Edger'	sieboldiana & forms
H. 'Gold Regal'	H. 'Sum and Substance'
H. 'Green Wedge'	The Tardiana group
H. 'Invincible'	H. tardiflora
H. rupifraga	H. tokudama & forms
H. 'Sea Drift'	H. 'Zounds'

Thrips

Small, cylindrical insects which can be a problem to hostas in hot, dry summers and also to hostas grown under glass. They can be controlled with contact insecticides such as Malathion, Nicotine and HCH.

Vine Weevils

The larvae should always be picked out and crushed underfoot when discovered. Effective control can be achieved for hostas grown in pots by mixing HCH powder into the potting compost, and top-dressing the pots with HCH again every few months. Control is less easy in the garden, but the use of HCH powder worked into the soil round affected plants is generally successful.

Rabbits

Rabbit populations in Britain are increasing and the resulting pressure means that they are increasingly invading gardens in their search for food. The damage they do can be considerable and heart-breaking and in areas where rabbits are a real problem, the answer seems to be to grow only rabbbit-proof plants. Hostas, unfortunately, are not rabbit proof.

Close-mesh wire fencing is effective in keeping rabbits out if the bottom 12 in (30 cm) is buried with the lower 6 in (15 cm) bent outwards to stop the rabbits tunnelling, but the cost is prohibitive. Shooting and gassing are both effective but should be done by professionals and, in any case, are impractical in small gardens. Traps and snares can also be effective but there are strict legal controls on these and, again, they are a danger to domestic pets.

Deer

In Britain the populations of deer are increasing and as with rabbits, this is causing them to become more of a garden problem. The only effective control is fencing, which must be at least 6 ft (1.8 m) high.

Diseases

So rarely are hostas troubled by disease that there is very little published information available on this subject. However disease is always more likely to strike an under-nourished individual struggling for survival than a healthy one. Thus good cultivation and the prevention as far as possible of damage by slugs and snails is essential to the prevention of disease.

Viral infections

Glaucous-leafed hostas of heavyish substance, and particularly some of the Tardiana group are more prone to virus infection than others, although *H. crispula* becomes badly affected especially if exposed to raw sewage. The Royal Horticultural Society's plant pathologist states that of the infected hostas in the National Reference Collection at Wisley Garden, the two viruses isolated for them by the Glasshouse Crops Research Institute, are Arabis Mosaic Virus, which produces chlorotic leaves with some stunting, and Tobacco Rattle Virus, which produces general chlorosis, and chlorotic ring spot systems, and stunted plants. This stunting and deformation of the leaves appears as excessive proliferation of one or two leaves in a clump, with the remainder stunted; or crinkling, crumpling, curling or rolling in the leaves. Sometimes mineral deficiency or chromosome deficiency show up in leaves as blotches or mottling and this is sometimes mistaken for virus infection.

Often virused leaves have a somewhat variegated appearance and this can be mistaken for a newly sported variegation. If such a virused hosta is propagated vegetatively, the likelihood is that all the propagules will also be virused, and that the virus will spread from garden to garden. Such plants should be burned before the virus spreads. It is simply not worth trying to cure and resuscitate a virused hosta, not even an expensive or rare one.

Viruses can also exist in a latent form in a hosta and will only manifest themselves if the plant is damaged in some way or divided to make other plants.

Crown rot

In Britain this is hardly a problem and indeed need not be a problem at all if good cultural procedures are followed. It is rarely, if ever, caused by bacterial pathogens. The cause is usually fungal, most frequently *Botrytis cinera*, the grey mould fungus which usually produces fluffy growth at the base of the crown. This problem usually first becomes visible when leaves turn yellow and are easily pulled off at the crown. The damaged area, which will emit a rotten smell, should be carefully cut away and the remainder of the hosta lifted and treated with a fungicide at the dosage recommended by the manufacturers. An American Hosta Society grower recommends immersing the remaining healthy crowns of the affected plant for eight hours in a plastic bucket containing the fungicide solution, making sure the roots are completely covered. The hosta should then be allowed to drain in the open, preferably in gently moving air, and then re-planted in sterile potting compost in new plastic pots, with the top of the crown slightly above the soil surface. The pot should be kept isolated from other hostas until the following season. Any remaining fungicide solution should be poured into the hole in which the infected hosta was growing. At regular intervals water the treated hosta with fungicide solution made up to the same strength as that used when first treating the infection.

In America the fungi *Corticum rolfsii* and *Schlerotium delphinii* cause crown rot. Hotter temperatures in parts of the southern states of America are responsible for the presence of the fungi *Fusarium oxysporum* and *Verticillium albo atrum* which cause progressive yellowing of the outer to the inner leaves of the hosta clump.

Crown rot can be caused by over-watering or too dense mulches packed tightly round the clump preventing movement of air or bad soil drainage. Hostas with short petioles and tightly packed leaves in compact clumps are most susceptible, *H.* 'County Park' being particularly affected. Hostas are normally very tough and can survive the rigours of packaging and transport but they should never remain wrapped in polythene or plastic for longer periods than is absolutely necessary.

Leaf spot

Leaf spot in hostas is caused by the fungi *Alternaria, Phyllosticta* and *Colletotrichum omnivorum*. However, it must also be pointed out that spots on hosta leaves can be caused by early frost damage, damage from misdirected chemical garden sprays and other air-borne pollutants. As with viral infection, leaf spot damage can be mistaken for the annual decay of the leaves in late autumn which causes similar spotting. Leaf scorch, similar to leaf spotting, is harmless, though unsightly, and is caused by small drops of cold water falling on to the leaves of hostas growing in strong sunlight.

The treatment of leaf spot is two or three sprays at fortnightly intervals with Benomyl or Thiram.

BIBLIOGRAPHY

ADEN, P., 1988: *The Hosta Book*, Timber Press, Portland, Oregon.

ALEXANDER, J.C.M., 1986: *European Garden Flora*, Vol. I, Cambridge.

AMERICAN HOSTA SOCIETY, *Bulletins* 1969–85, Journals 1986–7.

ANDREWS, H. 1797: 'New & Rare Plants', *Botanist's Repository*, p. 311, table 6.

ANONYMUS, 1863: Royal Horticultural Soc., 1891a 'Sempervirens 20', pp. 388–9, 391. 1891b 'Sempervirens 20'; pp. 531–3, 535.

ARENDS, G. 1951: 'Mein Leben als Gärtner und Süchter' ('My Life as Nurseryman and Breeder') – Grundlagen u. Fortschritte im Garten, u. Weinbau, Heft 91, Ludwigsberg. (Hosta p. 138.)

ASCHERSON, P. & GRAEBNER, P., 1905–7: *Synopsis der mittel-europaischen Flora*, III. – Leipzig. (Hosta p. 53–5.)

BAILEY, L.H., 1923: 'Various Cultigens, and Transfers in Nomenclature', Gentes Herbarum, I: 3, Ithaca, N.Y., pp. 113–36.

BAILEY, L.H., 1930: 'Hosta: the Plantain Lilies', 2: 3, op.cit. pp. 117–42.

BAILEY, L.H., 'Addenda' in Vol. II, particularly in relation to nomenclature, 2: 7, op.cit., pp. 425–41. (Hosta pp. 433–8.)

BAILEY, L.H., 1949: *Manual of Cultivated Plants, Revised Ed.*, New York. (Hosta p. 206–7.)

BAKER, J.G., 1868: 'The Genus Funkia' – *Gardeners' Chronicle*, London, p. 1016.

BAKER, J.G., 1870: 'A Revision of the Genera and Species of Herbaceous Capsular Camophyllous Liliaeceae', – *Journal of the Linnaean Society of London* II, nos. 54–5, pp. 349–436. (Funkia p. 366–8.)

BAKER, J.G., 1876: '*Funkia fortunei*, Baker', *Gardener's Chronicle*, London, N.S., 6, p. 36.

BANKS, J., 1791: *Icones selectae plantarum, quas in Japonica collegit et delineavit Engelbertus Kaempfer*, London. (*Hemerocallis japonica* p. 1, no. 11, table 11.)

BLOOM, A., 1974: *Perennials for your Garden*, Nottingham.

BRETSCHNEIDER, E., 1898: *History of European botanical discoveries in China*, London.

BRICKELL, C.D., 1968: 'Notes from Wisley', *Journal of the Royal Horticultural Society*, 93, pp. 365–72.

BRITISH HOSTA & HEMEROCALLIS SOCIETY *Bulletins*, 1983–7.

BOWERS, J.Z., 1970: *Western Medical Pioneers in Feudal Japan*, Baltimore & London.

CHUNG, M., 1989: *Annals of the Missouri Botanic Garden*, Vol. 76, no. 2, pp. 602–4.

Dictionary of National Biography, 1891: (Hibbert) p. 343, London.

DAHLGREN, R.M.T., CLIFFORD, H.T. & YEO, P.F., 1985: *The Family of the Monocotyledons*, Berlin & Heidelberg.

EMBERTON; S., 1968: *Garden Foliage for Flower Arrangement*, London.

ENGLER, A. 1888: *'Liliaceae'*. In: ENGLER, A. & PRANTL, K.: *Die natürlichen Pflanzenfamilien*, II Teil, 3. Abt., Leipzig. (Hosta pp. 39–40.)

FARRER, R., 1919: *The English Rock Garden*.

FRANCHET ET SAVATIER, 1879: *Enumerato Plantarum Japonicum*, II, p. 529.

FUJITA, N., 1976a: 'The Genus Hosta (Lilaceae) in Japan', *Acta Phytotax. Geobotanica* Vol. 27 (3–4), pp. 66–96. (Translated in *AHS Bulletin* Vol. 10, pp. 14–43 (1978).

Garden & Forest, 1893: unattributed (Thomas Hogg).

GRENFELL, D., 1981: 'A Survey of Hosta and its availability in Commerce', *The Plantsman* 3, pp. 20–44

GRENFELL, D., 1982: 'Hostas: The Golden Age', *Journal of the RHS*, 107: pp. 195–9.

GREY, Ch. H., 1938: 'Hardy Bulbs ...', *Liliaceae*, London. (Hosta pp. 292–305.)

HAMBLIN, Prof. S.F., 1936: *Lexington Leaflets*, Lex. (Mass.) Botanic Gdn., New York, pp. 211–24.

HEDRICK, U.P., 1950: *History of Horticulture in America in 1860*, New York.

HENSEN, K.J.W., 1963a: 'De in Nederland getweeke Hostas', ('Hostas cultivated in The Netherlands'). *Mededelingen van de Directeur van de Tuinbouw*, Vol. 26, pp. 725–35.

HENSEN, K.J.W., 1963b: 'Identification of the Hostas ('Funkias') introduced and cultivated by von Siebold', *Mededelingen Landbouwhogeschool, Wageningen*, Vol. 63(6), pp. 1–22.

HENSEN, K.J.W., 1963c: 'Preliminary registration lists of cultivar names in Hosta Tratt.', *Mededelingen Landbouwhogeschool, Wageningen*, Vol. 63, part 7, pp. 1–12.

HENSEN, K.J.W., 1983: 'Some nomenclatural problems in Hosta', *A.H.S. Bulletin*, Vol. 14, pp. 41–4.

HENSEN, K.J.W., 1985: 'A Study of the Taxonomy of Cultivated Hostas', *The Plantsman*, Vol. 7, part 1.

HOOKER, W.J., 1838, and 1839: *Curtis' Botanical Magazine*.

HOUTTUYN, 1780: *Natuurliche Historie*, 'Deel II' (1773–85), Amsterdam.

HUXLEY, A.J., 1981: *The Penguin Encyclopedia of Gardening*, London.

HYLANDER, N., *Lustgarden*, 1953.

HYLANDER, N., 1954: 'The Genus Hosta in Swedish Gardens', *Acta Horti Bergiani*, Vol 16, no. 11, pp. 331–420.

HYLANDER, N., 1954: 'The Genus Hosta', *Journal of the RHS*, 85, pp. 356–66.

ICBN = INTERNATIONAL CODE OF BOTANICAL NOMENCLATURE – *Regnum Vegetabile*, Vol. 3, Utrecht, 1952.

IINUMA, Y., 1874: *Somoku Dzesetsu*, 1:6., Tokyo (funkia tab. 19–27).

IR = INTERNATIONAL RULES OF BOTANICAL NOMENCLATURE, Ed. 3 – Jena 1935.

IRVING, W., 1903: 'The Funkias', *The Garden* 64, London, p. 297.

JEKYLL, G., 1908: *Colour Schemes for the Flower Garden*, Woodbridge.

KAEMPFER, E., 1712: *Amoenitatum Exoticarum Politico-Physico-Medicarum Fasciculi V.*, Fasc. 5 – Lemgo (Germany).

KER-GAWLER, J., 1812: '*Hemerocallis japonica*, Sweet-scented Day-Lily of Japan', *Curtis' Botanical Magazine*, 35, tab. 1433.

KOIDZUMI, G., 1930: *The Botanical Magazine*, 44, Tokyo, 514.

KORNER, H., 1967: *Die Wirzburger Siebold, eine Gelehrtenfamilie*, des 18. und 19, Jahrhunderts. (Lebensdarstellung deutscher Naturforscher herausgegeben von der Deutschen Akademie der Naturforscher Leopoldina, Nr. 13.)

KUNTH, C.S., 1843: *Enumeratio Plantarum omnium huisque cogniaraum . . . IV.*, Stuttgart & Tubingen. (Funkia pp. 590–2.)

LACEY, A., 1986: 'Hosta Revival', *Horticulture*, pp. 28–34.

LANNING ROPER, 1959: *The Gardens in the Royal Park at Windsor*, London.

LINDLEY, J., *Edwards Botanical Register*, 1839, Reg. 35, pl. 50.

LLOYD, C., 1973: *Foliage Plants*, London.

LLOYD, C., 1984: *The Well-Chosen Garden*, London.

LODDIGES, C., 1982: *Botanical Cabinet*, 19, 1869.

MAEKAWA, F., 1940: 'The Genus Hosta', *Journal of the Faculty of Science, Imperial University Tokyo*, Section III Botany, Vol. 5, pp. 317–425. (Re-printed in *AHS Bulletin*, Vol. 4, pp. 12–64 (1972) and 5, pp. 12–59 (1973).

MAEKAWA, F., *New Flowering Plants*, 70, pp. 3–7, Tokyo.

MATSUMURA, J., 1905: *Index plantarum japonica*, II – Tokioni. (Hosta pp. 199–201.)

MERRILL, E.D., 1938: 'A Critical Consideration of Houttuyn's new Genera and new Species of Plants' *Journal of the Arnold Arboretum*. 19:4, pp. 291–375 (Hosta p. 324).

MIQUEL, F.A.W., 1867: 'Prolusio Florae Iaponicae, V., *Ann. Mus. Lugduno-Batavi*, 3, pp. 91–209. (Funkia p. 152–3.)

MIQUEL, F.A.W., 1869a: 'Bijdragen tot de Flora van Japan, I. Funkia Spr.', Verslad *Mededelingen Akademie Wetensshappen Amsterdam*, Natuurkunde Series, Vols. 2 & 3, pp. 295–305. (Translated into French in *Arch. neerl. Sci. exact. nat.*, Vol. 4, pp. 219–30 (1869)).

MOTTET, S., 1897: 'Les *Funkia*', *Revue hort.*, Paris, 69, pp. 114–16.

MURRAY, J.A., 1784: 'Caroli a Linne . . . Systema vegetabilium . . .', Ed. 14-Gottingae.

NASH, G.K., 1911: The Funkias or Day-Lilies, Torreya, 11: 1, pp. 1–9, New York.

DE NOTER, R., 1905: 'Les *Hemerocallis* et *Funkia*', *Rev. hort.*, Paris, 77 (N.S., 5), pp. 388–91. (Funkia pp. 390–1)

OHWI, J., 1965: 'Flora of Japan', Washington, DC, pp. 287–91.

OTTO, F. & DIETRICH, A., 1833: *Cultur und Beschreibung der Funkia undulata Nob. Allgeneine Gartenzeitung*, Vol. 1, p. 119.

PAXTON, J., 1838: *Magazine of Botany* 5, pp. 25–6.

REDOUTÉ, P.J., 1802: *Les Liliaceaes*, I., Paris.

REGEL, E., 1876: 'Die *Funkia*-Arten der Garten und Deren Formen', *Gartenflora*, Stuttgart, 25, pp. 161–3.

ROBINSON, W., 1883: *The English Flower Garden*, London, p. 518.

ROYAL HORTICULTURAL SOCIETY, 1969: *RHS Dictionary of Gardening, Supplement, Second Ed.*, 'Hosta', (Hylander), pp. 347–50.

RUH, P., 1980: *Preliminary checklist of genus Hosta*, Am. Hosta Soc.

SALISBURY, R.A., 1807: 'Observations on the Perigynous Insertion of the Stamina of Plants', *Transactions of the Linnean Soc. of Lond.*, 8, pp. 1–16. (Saussurea p. 11).

SALISBURY, R.A., 1812: 'On the cultivation of Rare Plants, especially such as have been introduced since the death of Mr. Philip Miller', *Transactions of the Horticultural Soc., Lond. 1* (Niobe, Bryocles, p. 335).

SILVA TAROUCA, E., 1910: *Unsere Frieland-Stauden*, Leipzig & Vienna (Funkia p. 103.).

SIEBOLD & ZUCCHARINI, *Flora Japonica*, 1835.

SPRENGEL, K., 1817: 'Anleitung zur Kenntniss der Gewachse, 2.' *Ausg.*, II: 1. – Halle. (Funkia p. 246).

SPRENGEL, K., 1825: *Caroli Linnaei ... Systema Vegetabilium*, Ed. 16a, II, Gottingen. (Funkia pp. 40–1.)

STEARN, W.T., 1931a: '*Lilium cordatum* and *Hosta japonica.* Two Thunbergian Species', *Gard. Chron.*, London, 3 Ser., 89, p. 111.

STEARN, W.T., 1931b: 'The Hostas or Funkias. A revision of the Plantain Lilies', Ib., 90, pp. 27, 47–9, 88–9, 110.

STEARN, W.T., 1932a: in *Notes from the University Herbarium*, Cambridge, 3. 'Schedae ad Sertum Cantabrigiense Exsiccatum a W.T. Stearn editum'. Decades I-II. – *Journal of Botany*, London, 70, Suppl., pp. 1–29. (*Hosta lancifolia* p. 1415.)

STEARN, W.T., 1932b: '*Hosta (Funkia) rectifolia.* A newly identified Plantain Lily', *Gard. Chron.*, London, 3 Ser., 91, p. 23.

STEARN, W.T., 1947: 'The Name *Lilium japonicum* as used by Houttuyn and Thunberg', *RHS Lily Year Book*, London, 11, pp. 101–8.

STEARN, W.T., 1948: 'Kaempfer and the Lilies of Japan', Ib., 12, pp. 65–70.

STEARN, W.T., 1951: '*Hosta rectifolia*' *Curtis's Bot. Mag.*, 168, tab. 138.

STEARN, W.T., 1953: '*Hosta tardiflora*', Ib. 169, tab. 204.

STEARN, W.T., 1971: 'Philip Franz von Siebold and the Lilies of Japan', *RHS Lily Year Book*, London, 34, pp. 11–20.

THOMAS, G.S., 1960: 'Notes on the Genus Hosta', *Hardy Plant Society Bulletin*, Vol. 2, No. 5, pp. 92–101.

THOMAS, G.S., 1970: *Plants for Ground Cover*, London.

THOMPSON, P., 1980: 'Hostas from Seed', *Jour. of the RHS*, 105, pp. 371–2.

THUNBERG, C.P., 1784: 'Flora Japonica', (Leipzig)., 1780: 'Kaempferus Illustratus' (1), Nova Acta Soc. Sci. Upsal., 3.

TRATTINICK, L., 1812: *Archiv der Gewachskunde, I.*, Vienna. (Hosta p. 55–6.)

TREHANE, P., 1989: *Index Hortensis, Vol. 1: Perennials*, Wimborne, Dorset.

WALLACE, A., 1906: 'Scots Gardeners in America', *The Journal of Horticulture*.

WATANABE, K., 1985: *The Observation and Cultivation of Hosta*, New Science, Tokyo.

WEHRHAHN, H.R., 1931: *Die Gartenstauden, I*, Berlin & Altenburg. (Hosta pp. 66–8.)

WEHRHAHN, H.R., 1934: 'Die Funkien unserer Garten', *Gartenschönheit*, Berlin-Westend, 15, pp. 180–1, 203–4.

WEHRHAHN, H.R., 1936: 'Zwei ubersehene *Hosta*-Arten', *Gartenflora*, Berlin, 85, pp. 246–9.

WHERRY, E.T., 1946: 'A Key to the Cultivated Hostas', – *National Horticultural Magazine*, Baltimore, Md., 25: 3, pp. 253–6.

DE WOLF, G., 1982: editorial, *Horticulture*, p. 5. (information on Thomas Hogg)

WRIGHT, C.H., *The Botanical Magazine*, 1916, l. 8645.

HOSTA SUPPLIERS

United Kingdom

Apple Court,
Hordle Lane,
Lymington,
Hampshire
SO41 0HU

Ann & Roger Bowden,
Cleave House,
Sticklepath,
nr. Okehampton,
Devon
EX20 2NN

Bressingham Gardens,
Diss,
Norfolk
IP22 2AB

Goldbrook Plants,
Hoxne,
Eye,
Suffolk
IP21 5AN

Hadspen House,
Castle Cary,
Somerset
BA7 7NG

Kittoch Plants,
Kittoch Mill,
Carmunnock,
Glasgow
G76 9BJ

Mallorn Gardens,
Lanner Hill,
Redruth,
Cornwall
TR16 6DA

Park Green Nurseries,
Park Green,
Wetheringsett,
Stowmarket,
Suffolk
IP14 5QH

Europe

Heinz Klose,
3503 Lohfelden bei Kassel,
Rosenstrasse 10,
W. Germany

Staudengartnerei Grafin von
Zeppelin,
D-7811,
Sulzburg-Laufen,
W. Germany

Dr Hans Simon,
D-8772,
Marktheidenfeld,
W. Germany

Ignace van Doorslaer,
De Ten Ryen,
Kapellendries 52,
B-9231 Melle Gontrode,
Belgium

Arie van Fliet,
C. Klyn & Company,
Zuidkade 101,
2771DS, Boskoop,
Holland

United States

Pauline Banyai,
626 West Lincoln Avenue,
Madison Heights, MI 48071
USA

Busse Gardens,
635 E. 7th Street,
Cokato, MN 55321
USA

Coastal Gardens,
4611 Socastee Blvd.,
Highway 707,
Myrtle Beach, SC 29575
USA

Englerth Gardens,
2461 22nd Street, Rt. 2,
Hopkins, MI 49328
USA

Homestead Division,
944 8 Mayfield Road,
Chesterland, OH 44026
USA

Klehm Nursery,
Route 5, Box 197,
South Barrington, IL 60010
USA

Powells Gardens,
Route 3, Box 21,
Princeton, NC 27569
USA

Savory's Greenhouses &
Gardens,
5300 Whiting Avenue,
Edina, MN 55435
USA

André Viette Nurseries,
Route 1, Box 6,
Fisherville, VA 22939
USA

Japan

Kamo Nurseries,
Kami-Marubuchi,
Sofue Cho, Nakajima,
Aichi Prefecture
Japan

Kenji Watanabe,
59–1 Nagatsuka,
Gotemba-shi,
Shizuoka-Ken
Japan

The majority of suppliers listed are small specialist nurseries concentrating on hostas and some may not offer a postal service. This should be established when writing for a catalogue.

GENERAL INDEX

INDEX OF HOSTA NAMES

Figures in bold indicate main text references; in italic illustration references.
Colour plates are between p.128 and p.129.